Nuclear Fission - From Fundamentals to Applications

Edited by Pavel Tsvetkov

Published in London, United Kingdom

Nuclear Fission - From Fundamentals to Applications
http://dx.doi.org/10.5772/intechopen.104059
Edited by Pavel Tsvetkov

Contributors
Ernie Kee, Martin Wortman, Valerie Pelton, Modesto Montoya, Saurabh Sharma, Masako Ikegami, Ahmed Shaker, Eberhard H. Lehmann, Anna C. Hayes, Pavel Tsvetkov, Thabit Abuqudaira, Piyush Sabharwall

First published in London, United Kingdom, 2024 by IntechOpen
IntechOpen is the global imprint of INTECHOPEN LIMITED, registered in England and Wales, registration number: 11086078, 5 Princes Gate Court, London, SW7 2QJ, United Kingdom

British Library Cataloguing-in-Publication Data
A catalogue record for this book is available from the British Library

Additional hard and PDF copies can be obtained from orders@intechopen.com

Nuclear Fission - From Fundamentals to Applications
Edited by Pavel Tsvetkov
p. cm.
Print ISBN 978-1-83768-769-5
Online ISBN 978-1-83768-770-1
eBook (PDF) ISBN 978-1-83768-771-8

For EU product safety concerns:
IN TECH d.o.o., Prolaz Marije Krucifikse Kozulić 3, 51000 Rijeka, Croatia,
info@intechopen.com or visit our website at intechopen.com.

Meet the editor

Pavel V. Tsvetkov, Ph.D., is a professor in the Department of Nuclear Engineering, Texas A&M University, USA. Dr. Tsvetkov's research program is focused on novel energy systems meeting global growing needs for sustainable resources. His project portfolio includes direct energy conversion, waste minimization efforts, novel reactor designs, instrumentation efforts, and data science and engineering for a broad range of applications targeting optimized designs and performance. Dr. Tsvetkov is a member of the American Nuclear Society, American Society of Mechanical Engineers, Institute of Nuclear Materials Management American Institute of Chemical Engineers, American Society of Engineering Education, Alpha Nu Sigma, and Phi Kappa Phi. He has published more than 300 papers in peer-reviewed journals, conference proceedings, and reports as well as served as an editor and major contributor for fourteen books on energy, environment, and nuclear energy.

Contents

Preface

This book presents nuclear fission as a contemporary clean source of sustainable energy. Readers are taken on a journey that begins with policy and socioeconomic questions, continues toward technology considerations, and concludes with examples of applications. The idea is to account for the full range of relevant considerations. This book includes the fundamental principles, technological advancements, policy implications, and innovative applications that define the field of nuclear fission. From nuclear physics to global energy policies, this exploration seeks to provide readers with information on science and socio-political complexities surrounding nuclear fission as an energy source.

Nuclear fission, with its origins in the early 20th century, has evolved from being a scientific subject to becoming one of the key energy sources of the contemporary energy landscape. As we navigate the challenges of a growing global population, a limitless demand for energy, and growing environmental challenges, harnessing nuclear fission for practical applications becomes paramount.

The story of nuclear fission extends beyond the confines of laboratories and power plants. The book begins with policy and economics examples that showcase the development and deployment of nuclear technology. Beyond the laboratory, nuclear fission has become a critical player in the global energy landscape. The book explores the deployment of nuclear power, examining the technological advancements that have made it a viable and efficient source of energy. Delving into the economic, environmental, and geopolitical considerations, readers will gain insight into the role nuclear power plays in the broader context of sustainable energy solutions.

Ultimately, the journey proceeds with a dive into the fundamentals of nuclear fission while relating to such practical applications as novel sensors. Nuclear fission continues to inspire innovation. Novel sensors and technologies are emerging, pushing the boundaries of what is possible. This book examines the latest advancements and needs, from extreme conditions found in advanced reactor environments to the development of sophisticated monitoring systems, showcasing ongoing efforts to enhance safety, efficiency, and environmental sustainability.

This book endeavors to offer readers a nuanced perspective on the science that underpins nuclear fission and its far-reaching applications. As we stand at the crossroads of scientific and societal progress, the significance of nuclear fission applications cannot be overstated. This book aims to provide readers with the information that will inspire their interest in navigating the complex terrain of nuclear science and its multifaceted impact on our world. Whether you are a student, a researcher, a policy maker, or an interested citizen, this exploration invites you to engage in a range of contemporary topics related to nuclear fission and its practical uses — from its fundamental principles to its diverse applications—and to contemplate its role in shaping the future of our global energy landscape.

The book examines the design, development, and commercialization of nuclear reactors with a focus on contemporary and emerging novel technologies. Policy and deployment considerations, technologies and applications are also included, as are questions of economic competitiveness and public acceptance. Readers of this book are provided with coverage of nuclear fission and its benefits as a clean and reliable energy source while extending into such topics as nuclear physics and the status of sensors for nuclear energy applications.

Pavel Tsvetkov
Department of Nuclear Engineering,
Texas A&M University,
College Station, Texas, United States

Section 1

Policy and Economics

Chapter 1

Perspective Chapter: The Cost of Energy Independence

Valerie Pelton

Abstract

To reduce potential exposure to economic fluctuations and dependence on extra-territorial energy sources, some European Union (EU) member states are exploring energy diversification as a path to energy independence, economic vitality and national security. Nuclear power is a leading alternative energy source which may enable the EU to weather and potentially avert energy insecurity crises, including ones arising out of or related to political crises, collective action by oil-producing nations, global oil demand, and oil price fluctuations. Although nuclear power may be a long-term energy solution, the risks and cost trajectories deter many EU member states from nuclear power plant construction due to operation, maintenance and security costs. This chapter considers some key costs and potential options to reduce costs.

Keywords: cost, discount rate, price, supply, production, coal, gas, uranium, EU, OPEC, IMF, Euratom Treaty, Nord Stream, power plant, power generation, renewables, reactor

1. Introduction

Although global energy markets began tightening in 2021 as economies emerging from the COVID-19 pandemic unleashed pent-up demand in manufacturing, transportation, and travel and as consumers gradually resumed pre-pandemic activities, [1] the global energy market contraction accelerated in 2022 with the disruption of supply chains and the rise in energy prices caused by uncertainty over energy supplies [2] and weather-related factors [3].

Due in large part to the cuts in oil supplies by Russia [4] after its invasion of Ukraine on February 24, 2022, [5] and lowered production targets by the Organization of Petroleum Exporting Countries [6] (OPEC), this contraction in global energy supply fueled rising inflation rates worldwide. The emerging energy crisis deepened in September 2022 after Russia cut off oil supplies to Europe from the Nord Stream 1 pipeline [7]. In November 2022, OPEC further cut production in response to the weakening global economy, rising interest rates and growing fear of a potential global recession [8].

Against this backdrop, the International Monetary Fund (IMF) cut projected global growth rate to 3.2% for 2022 and to 2.7% for 2023, a significant decrease from the 6% achieved in 2021 [9]. The IMF identified energy as a key driver of global inflation in **Figure 1** in its October 2022 *World Economic Outlook* report, as illustrated below ([9], at 4).

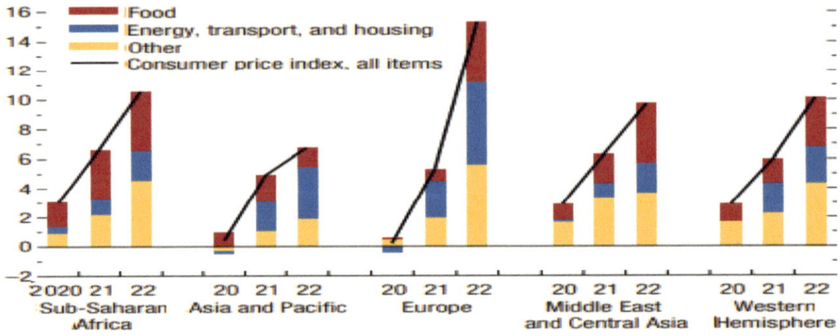

Figure 1.
Inflation driven by food and fuel (Annualized percent). Sources: IMF, Consumer Price Index database; and IMF staff calculations. Note: Figure shows inflation contributions from broad categories. Contributions are computed first by country, annualized over available months in cases in which data are partial (for example, for 2022). The figure shows both the median contributions and aggregate inflation rate for each region.

In response to unprecedented high fossil fuel costs which "account [ed] for 90% of the rise in the average costs of electricity generation worldwide (natural gas alone [accounted] for more than 50% [of the increase])", ([10], at 34) EU member states began making difficult energy-related economic decisions in 2022. Exacerbating the energy crisis, EU member states imposed sanctions on Russia, including cessation of coal imports ([10], at 33) As of August 2022, coal "deliveries from Europe's largest external supplier fell to zero" [10]. Prior to Nord Stream 1 pipeline supply cut-off, Russian energy deliveries to the EU were down by 80% by September 2022 ([10], at 33). As the war in Ukraine continues and constricted access to energy resources continues to drive inflationary costs and pricing, EU members states continue to grapple with measures to alleviate the economic impact of Russia's cutting gas deliveries and the EU's ban on Russian coal importation.

2. Price vs. cost

2.1 Definitions

As the price and cost of energy resources influence decisions about energy resources, it is important to understand the terms are not an interchangeable. For purposes of this chapter, the term "price" is defined as

> *"a: the amount of money given or set as consideration for the sale of a specified thing*
>
> *b: the quantity of one thing that is exchanged or demanded in barter or sale for another" [11].*

and the term "cost" is defined as

> *"a: the amount or equivalent paid or charged for something*
>
> *b: the outlay or expenditure (as of effort or sacrifice) made to achieve an object" [12].*

2.2 Price

With respect to the price of uranium ore, the per pound price was $48.15 on December 16, 2022 [13]. In comparison, the price of crude oil on December 17, 2022 ranged from $74.29 (WTI crude) to $80.47 (OPEC basket) [14]. While the price of uranium ore is relatively low compared to that of crude oil, the price of uranium's fuel form increases once processed and enriched due to the costs of processing and enrichment. In its 2021 Uranium Marketing Annual Report, the U.S. Energy Information Administration (USEIA) noted civilian nuclear power reactor owners and operators (collectively, COOs) purchased a total of 46.7 million pounds of uranium (equivalent) at a weighted-average price of $33.91 per pound [15]. In comparison, as natural gas prices skyrocketed in 2021, the global demand for coal power generation increased 9% to "10,350 terawatt-hours (TWh)" ([16], pp. 8, 10). Although EU coal consumption was projected 45 metric tonnes (MT) in 2021, the price of coal quadrupled to $170/MT in September 2021 [16] and soared to an "all-time high of around $450 per tonne" while the price of uranium was $48.10 [17].

3. Cost

3.1 Processing and enrichment

In addition, COOs purchased 14 million separative work units (SWU) under enrichment services contracts from 11 sellers at an average price of $99.54 per SWU [18]. According to the WNA, the $1663 front-end fuel cycle costs of 1 kg of uranium as of September 2021 were significantly higher than for raw uranium ore ($842 for 8.9 kg U3O8×$94.6/kg) due to the cost of conversion($120 for 7.5 kg U×$16), enrichment ($402 for 7.3 SWU×$55) and fuel fabrication ($300) ([18, 19], p. 62).

In 2011, the World Nuclear Association ("WNA") concluded "about half of the cost [of nuclear power] is due to enrichment and fabrication" [17]. Although those costs are significant, the WNA concluded nuclear power is "cost competitive with other forms of electricity generation, *except where there is direct access to low-cost fossil fuels*" [emphasis added] [19]. However, the WNA also acknowledged "nuclear technologies are capital intensive relative to natural gas or coal, [and] the cost of nuclear rises relatively quickly as the discount rate is raised", [18] seemingly undercutting its 2022 assertion of nuclear power's cost competitiveness. However, after applying a 7% discount rate, "the median value of nuclear is close to the median value for coal ..., and at a 10% discount rate the median value for nuclear is higher than that of either CCGTs or coal" [18].

3.2 Old vs. new nuclear plants

While nuclear power generated by existing older plants may be competitive with fossil fueled plants, the unit cost of nuclear energy by new nuclear power plants which must amortize the costs of new construction are higher than for older nuclear and new fossil fuel power plants. In large part, the higher cost of power generated by new nuclear plants is due to finance costs and construction cost overruns [20]. Specifically, the high capital costs for plant construction, lengthy construction time, interest fees on capital used to finance construction and lengthy time period before a power plant begins generating for new nuclear power plant construction [21]. In addition, the cost of nuclear accidents is also a factor. If such costs were "allocated to the aggregate output of nuclear power plants . . . [in 2011, they] would [have led] to a price increase of 2.3 euro cent per kWh" [22]. EDF Group, one of the world's largest power companies, estimated the minimum construction cost of a new nuclear power plant in France will be €950 million (approximately €5700 per kW) before 2030 [23]. Consequently, nuclear plant construction typically requires some form of government subsidization/support ([24], p. 17).

While the cost to bring a new nuclear power plant online (e.g., fuel, construction, safety, accident liability, environmental impact, etc.) may be daunting, the levelized cost of electricity (LCOE) for "nuclear energy plants coming online in 2020 was $95.2/megawatt hour (MWh), comparable to conventional coal ($95.1/MWh), above conventional combined cycle natural gas-fired plants ($75.2/MWh) but below conventional combustion turbine natural gasfired plants ($141.5/MWh)" [24, 25]. In a 2020 joint study by the International Energy Agency (IEA), Nuclear Energy Agency (NEA) and Organization for Economic Co-Operation and Development (OECD), estimated the LCOE in U.S. dollars per MWh for nuclear plants built 2020–2025 based on specific discount rates and an assumed 85% capacity factor [26]. Depending on the discount rate applied and the country in which construction occurs, the projected LCOE per MWh of a new nuclear plant will vary. Accordingly, while the LCOE of new nuclear plant in France is estimated at $45.3 (3% discount rate), $71.1 (7% discount rate) and $96.9 (10% discount rate), the LCOE of a similar new plant in Slovakia is estimated at $57.6 (3% discount rate), $101.8 (7% discount rate) and $146.0 (10% discount rate) [27].

3.3 Cost effectiveness

Historically, the 1957 Treaty on the European Atomic Energy Community (Euratom Treaty), as amended, [27] was created in part to enable EU member states to develop civilian nuclear power as an inexpensive, plentiful source of power generation [28]. By 2019, the Euratom Treaty contributed to the EU receiving "25% of electricity ... from nuclear, 46% ... from fossil fuels and biomass, and 29% [from] renewables (including 11% from hydro)" [29]. As France is one of the EU's leading producers of nuclear energy, assuming its socio-economic approach of applying "standardized hypotheses (on GDP, demographics, public policies in other European countries, and electricity consumption)" is applicable to other EU member states, and assuming the majority of those states' respective "*existing nuclear units* were the most affordable ones" [emphasis added], nuclear power may be an economically tempting alternative to fossil fuels ([30], pp. 163–167).

In comparison to other renewable energy resources, the 2021 report commissioned by European Conservatives and Reformists (ECR) Group and Renew Europe concluded "nuclear energy is more cost-effective than [other] renewables ... and would remain the cheaper option in 2050" [30]. That conclusion may not hold true if uranium ore costs $100/kg (or more) as the resulting fuel cost is €5.50/MWh [30]. For example, after uranium prices peaked at $125–150/kg in 2008, they subsequently decreased to $100/kg by 2019 [31]. As the lion's share of nuclear power costs occur early on, its cost advantage over other renewables decreases if the weighted average cost of capital [31] increases. Accordingly, unless the direct and indirect costs of new construction can be lowered and the lengthy timeline from construction to power generation can be offset, fossil fuels may remain an economically attractive option.

As the EU's electricity generation is dependent on fossil fuel plants which rely heavily on fuel imports from Russia, ([32], pp. 206, 209) it makes economic sense for the EU to increase investment in alternative or supplementary energy sources and technologies such as nuclear power. Indeed, as early as 2018, the EU's use of renewable energy resources surpassed that of solid fossil fuels and accelerated in 2020 [33]. The EU's need to utilize renewables to provide a reliable and cost-effective supply of energy for civilian uses is underscored by its wide deployment and development in Russia and the People's Republic of China (PRC). In 2019, Russia and the PRC were at forefront of the "5.5 GW of new nuclear capacity" installed worldwide [34]. In the third quarter of 2022 EU energy product imports reached €70 billion, reflecting an 11.3% increase in petroleum imports and a 13.2% increase in natural gas imports from 2021 [35]. Although this increase may be attributed mainly to surging prices after Russia cutting off its exports of those commodities to the EU, the strain of filing the supply void of the 14.4% Russia supplied to the EU [36] until its late September 2022 cut-off underscores its sensitivity to global oil demand and oil price fluctuations – and the need for a stable, reliable and cost-effective energy supply within the EU.

4. Cost-reduction options

In November 2021, the French secretary of state for the digital sector, Cedric O, nuclear energy "is not an ideological question, it is a mathematical question" [37]. For EU member states to overcome cost hurdles posed by new construction and other related costs (e.g., regulatory oversight, waste water storage, spent fuel disposal, political opposition, public acceptance, insurance, environmental concerns, accident clean up, etc.), cost-reduction options are critical. As Europe imports 90% of its natural gas and 97% of its oil supplies, [38] technologies, techniques and design solutions may be the keys to efficient lower cost construction and decreased time to power generation in ways which do not compromise safety will be the drivers.

4.1 Cooperation

One seemingly straight-forward option is to increase cooperation within the EU on energy construction projects. Although this option may seem simple, multistate coordination among numerous government agencies is a complex process. Updates to the Euratom Treaty such as Council Directive 2014/87 [39] and the European Instrument for International Nuclear Safety Cooperation (effective on June 14, 2021, [40]), provide guidance and a framework for cooperation. Although the practical aspects of multistate cooperation such as equitable allocation of projects, fairness and open competition during the bid process, project management, regulatory oversight, national security, budgeting and cost management are harder nuts to crack, energy security is vital to EU economies and of paramount interest in light of the on-going hardships occasioned by the Nord Stream 1 pipeline cut-off, cessation of coal imports from Russia and production decrease by OPEC.

4.2 Plant size and design

In this winter of energy discontent, if the EU considers construction of new nuclear plants within its member states as a potential longer term solution to its energy woes, plant size and design must be factored in the planning and budgeting stages. Although smaller plants should require less materials and less time for construction than larger plants, small modular nuclear reactors (SMRs), "[d]ue to the loss of economies of scale, the decommissioning and waste management unit costs of SMR [is actually two to three times] higher than those of a large reactor" [41]. Such plants cost may cost over 30% "more than the traditional nuclear power plants by 2030s.. [as] each kWh of electricity would cost between 15 and 70% more ... due to economies of scale" [42]. While larger scale plants are more cost-effective from a power generation standpoint as the reactor typically produces 1000 MW compared to the 300 MW typically produced by a smaller reactor, [42], the "capital start-up costs are much lower.. [with fewer construction] cost risks", place less of a load strain on electricity grids, "can be manufactured and assembled at a central factory and then sent to the site of installation", [42]... "greater simplicity of design, economy of series production ..., short construction times, and reduced siting costs", potential for below ground level installation, "high level of passive or inherent safety in the event of malfunction" [43]. As there are multiple SMR manufacturers (e.g., GE Hitachi Nuclear Energy, NuScale Power, China National Nuclear Corporation, CAREM, SMART, Holtec International, Rosatom, etc.), [43] the EU may be able to drive better bargains on construction costs if it hosts open bids.

4.3 Management practices

According to a 2018 study by the Massachusetts Institute of Technology (MIT), mismanagement (i.e., poor construction management practices) is behind the large cost of nuclear plant construction [44]. As in any industry, benchmarking and utilizing best practices are tools to increase efficiency. In the case of nuclear power, MIT concluded mismanagement is the real reason why "cost [in the East] is $3000–$4000 per kilowatt, whereas in the West the cost is north of $8000 per kilowatt" when it benchmarked "failed projects and the successful projects" [45]. The study determined "indirect costs—those external to hardware—caused 72% of the cost increase ... Examples include rising expenditures on engineering services, on-site job supervision, and temporary construction facilities." [46]. Based on its review, the MIT study made the following cost management recommendations:

> *"Mak[e] sure that the design is complete, ... that you have fabricators and constructors on your design team early so that you know that what you design can be built."*
>
> *To minimize construction delays and cost over-runs, ensure "everyone has skin in the game, making sure the process can deal with and adapt quickly to change ..."*
>
> *To prevent supply chain gaps and/or delays, take steps to ensure the project has a "reliable supply of spare parts and trained workers" prior to starting construction." [47].*
>
> *Reduce materials used and "automate some construction tasks" [48].*

5. Conclusion

To achieve an economically acceptable and feasible level of independence from economic fluctuations and dependence on extra-territorial energy sources, the EU will need make a long-term investment in nuclear power to mitigate economic and national security risks of fossil fuel supply fluctuations. As the KWh cost of nuclear power is competitive with fossil fuel, safely reducing construction costs and the timeline to power generation are critical to widening development, deployment and utilization of nuclear power within the EU. By using benchmarking and best practices to drive efficiencies in indirect costs, the intensive capital costs of nuclear power construction may be reduced considerably and construction timelines decreased without adversely impacting safety.

Conflict of interest

The author declares no conflict of interest.

Notes/thanks/other declarations

The views expressed in this Article are the author's and do not necessarily represent the views of the BIED Society, the BIED Global Leadership Academy or the United States.

Author details

Valerie Pelton
BIED Society, USA

*Address all correspondence to: vjpelton@yahoo.com

References

[1] Global Energy Crisis. International Atomic Energy Agency (IAEA). 2022. Available from: https://www.iea.org/topics/global-energy-crisis. Hereafter, "Global Energy Crisis"

[2] Kolaczkowski M. How does the war in Ukraine affect oil prices? World Economic Forum. Mar. 4, 2022. Available from: https://www.weforum.org/agenda/2022/03/how-does-the-war-in-ukraine-affect-oil-prices/ [Hereafter, "Ukraine War Impact"]

[3] World Energy Outlook 2022. IAEA. 2022. Available from: https://iea.blob.core.windows.net/assets/830fe099-5530-48f2-a7c1-1f35d510983/WorldEnergyOutlook2022.pdf [Hereafter, "2022 Energy Outlook"]. p. 34. See also Global Energy Crisis. See supra note 1

[4] Ukraine War Impact. See supra note 2

[5] Ellyat H. Russian forces invade Ukraine. Reuters. Available from: https://www.cnbc.com/2022/02/24/russian-forces-invade-ukraine.html

[6] Lawler A. OPEC cuts oil demand growth forecast again as economic challenges mount. Reuters. Nov. 14, 2022. Available from: https://news.yahoo.com/opec-cuts-oil-demand-growth-121140721.html [Hereafter, "OPEC Demand Forecast"]

[7] Meredith S. Russia has cut off gas supplies to Europe indefinitely. Here's what you need to know. CNBC. Sept. 6, 2022. Available from: https://www.cnbc.com/2022/09/06/energy-crisis-why-has-russia-cut-off-gas-supplies-to-europe.html

[8] OPEC Demand Forecast. See supra note 5

[9] World Economic Outlook: Countering the Cost of Living Crisis. Oct. 11, 2022. p. xiii, xvi. International Monetary Fund (IMF). Available from: https://www.imf.org/en/Publications/WEO

[10] 2022 Energy Outlook. See supra note 3. pp. 36-37

[11] Available from: https://www.merriam-webster.com/dictionary/price [Accessed: 2022-12-17]

[12] Available from: https://www.merriam-webster.com/dictionary/cost [Accessed: 2022-12-17]

[13] Metal Spot Prices. Show the Planet, Inc. Dec. 16, 2022. Available from: https://www.dailymetalprice.com/metalpricecharts.php?c=u&u=lb&d=240#:~:text=%20%20%20%20Metal%20%20%20,12%2F16Dec%2016%2C%202022%20%2019%20more%20rows%20 [Accessed: 2022-12-17]

[14] Oil Price Charts. A Media Solutions trading as Oilprice.com. Dec. 17, 2022. Available from: https://oilprice.com/ [Accessed: 2022-12-17]

[15] 2021 Uranium Marketing Annual Report. U.S. Energy Information Administration. May 26, 2022. pp. 1, 45. Available from: https://www.eia.gov/uranium/marketing/pdf/2021%20UMAR.pdf

[16] Coal 2021: Analysis and forecast to 2024. IEA. pp. 6, 7. Available from: https://iea.blob.core.windows.net/assets/f1d724d4-a753-4336-9f6e-64679fa23bbf/Coal2021.pdf

[17] Trading Economics. Dec. 23, 2022. Available from: https://

tradingeconomics.com/commodity/coal [Accessed: 2022-12-26]

[18] Economics of Nuclear Power. WNA. Aug. 2022. Available from: https://world-nuclear.org/information-library/economic-aspects/economics-of-nuclear-power.aspx [Hereafter, “2022 WNA Economics”]

[19] Economics of Nuclear Power. WNA. Dec. 2011. Available from: https://world-nuclear.org/information-library/economic-aspects/economics-of-nuclear-power.aspx

[20] 2022 WNA Economics. See supra note 23

[21] Afman M, Benner J, Bles M, Blom M, Croezen H, Rooijers F, et al. Nuclear energy: The difference between costs and prices. CE Delft. July 2011. p. 14. Available from: https://cedelft.eu/wp-content/uploads/sites/2/2021/03/3475_FinalReportMaBMV.pdf [Hereafter, “Cost and Price”]

[22] Cost and Price. See supra note 28. p. 33

[23] Blom M, Rooijers F, Schepers B. Nuclear energy: The difference between costs and prices. CE Delft. Nov. 2011. Available from: https://cedelft.eu/publications/nuclear-energy-the-difference-between-costs-and-prices/

[24] 2022 WNA Economics. See supra note 18

[25] See also Lazard’s Levelized Cost of Energy Analysis — Version 14.0. Lazard. 2020. p. 2. Available from: https://www.lazard.com/perspective/lcoe2020

[26] Wang K. Nuclear Power Economics Cost. May 4, 2018. Physics 241 coursework. Stanford University, Citing: Zach Long. Economics of Nuclear Power as an Energy Source. Physics 241 coursework. Stanford University. Winter 2016. Available from: http://large.stanford.edu/courses/2018/ph241/wang-k2/

[27] Projected Costs of Generating Electricity. IEA, nuclear energy agency and organization for economic co-operation and development. 2020. Available from: https://iea.blob.core.windows.net/assets/ae17da3d-e8a5-4163-a3ec-2e6fb0b5677d/Projected-Costs-of-Generating-Electricity-2020.pdf

[28] Treaty on the European Atomic Energy Community. European Union. Mar. 3, 1957. Available from: https://eur-lex.europa.eu/eli/treaty/euratom_2016/oj

[29] Dreesen B. The European Union Must Keep Nuclear Power on the Table. Power Magazine. Oct. 1, 2020. Available from: https://www.powermag.com/the-european-union-must-keep-nuclear-power-on-the-table/ [Hereafter, “Keeping Nuclear Power”]

[30] Nuclear Power in the European Union. WNA. Oct. 2022. Available from: https://www.world-nuclear.org/information-library/country-profiles/others/european-union.aspx

[31] Katinka M, Brouwer LLM, Bergkamp L, editors. Road to EU Climate Neutrality by 2050: Spatial Requirements of Wind/Solar and Nuclear Energy and Their Respective Costs. Brussels, Belgium: ECR Group and Renew Europe, European Parliament. Jan. 2021. p. 205. Available from: https://roadtoclimateneutrality.eu/Energy_Study_Full.pdf

[32] The weighted average cost of capital consists of the risk free rate, government risk premium and project risk premium. See supra, note 31. p. 206, 209

[33] How Europe can cut natural gas imports from Russia significantly within a year. IAEA. Mar. 3, 2022

Available from: https://www.iea.org/news/how-europe-can-cut-natural-gas-imports-from-russia-significantly-within-a-year/, See A 10-Point Plan to Reduce the European Union's Reliance on Russian Natural Gas. IAEA. Mar. 3, 2022. Available from: https://iea.blob.core.windows.net/assets/1af70a5f-9059-47b4-a2dd-1b479918f3cb/A10PointPlantoReducetheEuropeanUnionsRelianceonRussianNaturalGas.pdf

[34] Energy statistics - an overview. Eurostat. Feb. 2022. Available from: https://ec.europa.eu/eurostat/statistics-explained/index.php?title=Energy_statistics_-_an_overview#:~:text=As%20for%20the%20structure%20of%20gross%20available%20energy,produced%20from%20coal%2C%20crude%20oil%20and%20natural%20gas

[35] Keeping Nuclear Power. See supra note 21

[36] EU imports of energy products - recent developments. Eurostat. Dec. 20, 2022. Available from: https://ec.europa.eu/eurostat/statistics-explained/index.php?title=EU_imports_of_energy_products_-_recent_developments#:~:text=The%20share%20of%20petroleum%20oils%20in%20the%20total,a%20result%20of%20higher%20prices%20of%20these%20commodities [Accessed: 2022-12-26]

[37] Amaro S. The road to COP26: Greenwashing or viable solution? Europe has a big decision to make on nuclear power. CNBC.com. Nov. 9, 2021 Available from: https://www.cnbc.com/2021/11/09/cop26-europe-has-a-big-decision-to-make-on-nuclear-power.html [Accessed: 2022-12-26]

[38] Kilic ZB. Europe turning to nuclear power amid global energy price increase: EU Commission due to announce its final list of green investments in a few months. Yeni Safak. Oct. 18, 2021. Available from: https://www.yenisafak.com/en/world/europe-turning-to-nuclear-power-amid-global-energy-price-increase-3582490 [Accessed: 2022-12-26]

[39] Council Directive 2014/87/Euratom of 8 July 2014 amending Directive 2009/71/Euratom establishing a Community framework for the nuclear safety of nuclear installations. Eur-lex. European Union. July 8, 2014. Available from: https://eur-lex.europa.eu/legal-content/EN/TXT/?uri=celex%3A32014L0087/

[40] Proposal for a Council Regulation establishing a European Instrument for Nuclear Safety complementing the Neighbourhood, Development and International Cooperation Instrument on the basis of the Euratom Treaty. European Commission. June 14, 2018. Available from: https://eur-lex.europa.eu/legal-content/EN/TXT/?uri=CELEX%3A52018PC0462&qid=1618826576514

[41] SMR Techno-Economic Assessment. Project 1: Comprehensive analysis and assessment, techno-economic assessment final report Vol. 1. United Kingdom Dept. of Energy and Climate Change. Atkins Limited and Ernst & Young LLP. July 21, 2016. Available from: https://www.gov.uk/government/publications/small-modular-reactors-techno-economic-assessment

[42] Small modular reactors – how do they compare to traditional nuclear reactors New York Energy Week. Energy Week Holdings, LLC, an EnerKnol, Inc. Company. Sept. 17, 2018. Available from: https://nyenergyweek.com/small-modular-reactors-compare-traditional-nuclear-reactors/#:~:text=A%20study%20found%20that%20power%20from%20small%20modular,nuclear%20power%20plants%2C%20due%20to%20economies%20of%20scale

[43] See Small Nuclear Power Reactors. WNA. May 2022. Available from: https://www.world-nuclear.org/information-library/nuclear-fuel-cycle/nuclear-power-reactors/small-nuclear-power-reactors.aspx

[44] McMahon J. 3 Reasons Nuclear Power Plants Are More Expensive In The West (It's Not Regulation). Forbes Magazine. Forbes.com. Oct. 1, 2018. Available from: https://www.forbes.com/sites/jeffmcmahon/2018/10/01/3-reasons-nuclear-reactors-are-more-expensive-in-the-west-hint-its-not-regulation/?sh=697977895d1a [Hereafter, "Why Western Plants Are Expensive"]

[45] Buongiorno J, et al. Nuclear energy in a carbon-constrained world: Big challenges and big opportunities. IEEE Power and Energy Magazine. Cambridge, Massachusetts, USA: Massachusetts Institute of Technology (MIT), The MIT Press; February 2019;**17**(2):69-72. © 2019 The Author(s). Available from: https://energy.mit.edu/wp-content/uploads/2018/09/The-Future-of-Nuclear-Energy-in-a-Carbon-Constrained-World.pdf

[46] Stauffer NW. Building nuclear power plants: Why do costs exceed projections? MIT Energy Initiative. Nov. 25, 2020. Available from: https://energy.mit.edu/news/building-nuclear-power-plants/ [Hereafter, Building Nuclear Plants]

[47] Why Western Plants Are Expensive. See supra note 44

[48] Building Nuclear Plants. See supa note 46

Chapter 2

Perspective Chapter: Study of Indian and Japanese Civilian Nuclear Cooperation and Their Cooperation Policies with Selected Countries, Including IAEA

Saurabh Sharma and Masako Ikegami

Abstract

The authors have reviewed the bilateral civilian nuclear cooperation for Indian and Japanese cases. The objective of this research is to understand the civil nuclear agreements and analyze the nuclear policy for both Indian and Japanese cases, as well as to measure the impact of the India-Japan bilateral nuclear agreement through the comprehensive review study. Firstly, the Indian case covers India's civilian nuclear cooperation with France, IAEA, Russia, UK, and the USA. Secondly, the Japanese case explores Japan's civilian nuclear cooperation with IAEA, the USA, EURATOM, Russia, and the UK. Lastly, India-Japan civil nuclear cooperation is discussed.

Keywords: civilian nuclear cooperation, India-Japan, nuclear policy, IAEA, bilateral

1. Introduction

In this chapter, a review of the case analyses of Indian civilian nuclear policies and Japanese civilian nuclear policies is conducted based on their bilateral civil nuclear cooperation and agreements with other countries, including the India-Japan civilian nuclear cooperation agreement. The objective of this research is to review various civil nuclear agreements to analyze the nuclear policy for both Indian and Japanese cases as well as to measure the impact of the India-Japan bilateral nuclear agreement. Thus, the Indian and Japanese Governments' agreement documents are reviewed and discussed in Sections 2 and 3 for India and Japan, respectively. Afterward, a detailed study of India-Japan bilateral cooperation and agreement is conducted in Section 4 emphasizing background, present status, and analysis. In Section 5, the summary and conclusion of this research have been described.

2. Indian case analysis

In this section, India's civilian nuclear policy is explained with respect to its civilian nuclear cooperation agreements with IAEA, USA, France, Russia, and UK.

2.1 India-IAEA civil nuclear cooperation

Agreement between the "Government of India and International Atomic Energy Agency (IAEA) for the Application of Safeguard to Civilian Nuclear Facilities" [1] was signed in Vienna on February, 2, 2009 [1]. This agreement has been reviewed in a detailed manner in the following parts:

Upon the mutual consensus, on one hand, India agreed not to use any item under the regulation of this agreement for the manufacturing of nuclear weapons or military purposes; similarly, on the other hand, IAEA would ensure the "safeguards" of items subject to this agreement [1]. The objective of these "safeguards" is to resist to withdrawal of safeguarded nuclear material from civilian use at any time and their application would expedite the multilateral and bilateral engagements, which are in partnership with India [1].

Items to which these safeguards are applicable are nuclear and nonnuclear material, equipment, technology, and any facility identified by India producing, using, processing, or storing nuclear material in the territory of India [1]. In order to implement, India would notify IAEA as a sovereign and provide a written declaration to IAEA stating its decision to place its civilian nuclear facilities under the safeguards of the IAEA [1]. India in its declaration shall affirm that the above-mentioned nuclear material and other components would be used solely for peaceful purposes and there would not be the manufacturing of any nuclear explosive device [1]. The cessation of these safeguards would be subject to the provisions of GOV/1621 1973 [1]. Therefore, nuclear would be no longer safeguarded under this deal if they returned to the state which provided them to India. Also, if it is decided by IAEA that nuclear material is neither usable nor placed at notified compound subject to IAEA safeguards. Further, about the transfer of nuclear material, this agreement states that unsafeguarded material would be transferred outside the territory of India only when the material is returned to the supplier state in agreement with the Agency [1]. The notification to transfer unsafeguarded nuclear material from India to other countries should be made in advance to IAEA to enable it to take required actions and make necessary arrangements [1].

The part third of this agreement is broadly about safeguard procedures mentioned in the agreement. It would be the responsibility of IAEA to include "India Subsidiary Arrangements" [1] for safeguards implementation in the following manner [1]:

- For safeguards' effective implementation IAEA would review the design of the principal nuclear facility in its initial phase and India must facilitate design review by providing relevant information to the Agency [1].

- Reserved at least for 2 years, India must keep records of nuclear material at the principal nuclear facility, and hence India and IAEA shall agree to maintain "system of record with respect to each facility and also with respect to such material" [1].

- India must prepare reports about the production and processing of above mention recorded nuclear material, various conditions at a nuclear facility, and submit it to IAEA. And, if IAEA requests clarification, then India must submit the required "Amplification of Reports" [1].

- To check the fulfillment of safeguards, IAEA may conduct an inspection under the regulatory framework of this agreement, such as the audit of reports and records, review of design and functioning nuclear facility, and examination of operations, being carried out at the principal nuclear facility [1].
- Based on mutual consensus minimum of two reports per year must be submitted by India to IAEA about the routine inspection of nuclear reactors and nuclear material at the principal nuclear facility and other associated sites [1].
- A "Reprocessing Plant" holding more than 5 kg effective annual nuclear material shall be inspected all throughout and a regular monthly report must be submitted from India to IAEA [1].

Discussing the Agency Inspectors, the next segment states that, as applicable to relevant provisions, Inspectors from the Agency would execute tasks pursuant to this agreement. Furthermore, as per IAEA recommendation document INFCIRC/225/Rev.4 [1], India is obliged to undertake all appropriate essential measures for the "physical protection" [1] of nuclear material and facilities under its territorial jurisdiction subject to this agreement. For effective safeguards implementation, India must maintain a "system of accounting and control" to keep a check on all elements subject to these safeguards. Financing of all activities under this deal would be bear by parties equally, except in case of a special arrangement that will be reimbursed by IAEA to India [1]. If any defiance to the agreement by India, is observed by IAEA Board, it can summon India to resolve such matters and submit reports of its remedial measures and if India fails to undertake measures in a given time, then IAEA Board may take further actions under Article XII. C of the Statute [1]. Both parties must cooperate for the effective implementation of the agreement and resolve their dispute through mutual negotiations and consultations. Upon completion of constitutional and statutory requirements by India, it will provide written notification to IAEA to make the agreement's entry into force. Once entered into force, an agreement can be amended on the request of either of the party.

In the last section of this agreement, below mentioned are the definitions of important terms in the context of the legal agreement document to avoid any vagueness (**Figures 1** and **2**) [1].

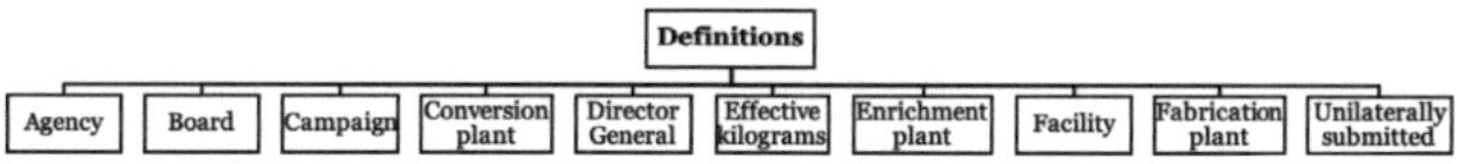

Figure 1.
Definitions 1 [1].

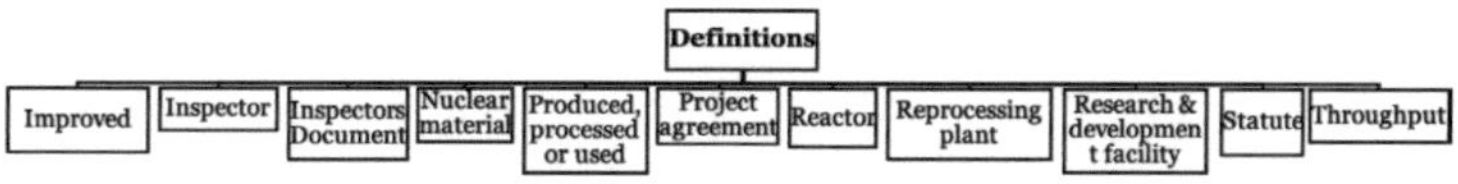

Figure 2.
Definitions 2 [1].

2.2 India-USA civil nuclear cooperation

To fulfill the objectives under this agreement, Article 1 [2] provides definitions of important terms used in this agreement (**Figures 3** and **4**). Article 2 [2] discusses the scope of cooperation about the usage and transfer of nuclear material and other associated components as per the provisions of this agreement. Areas of cooperation include advanced research on nuclear science, nuclear safety, collaborative research through workshops, meetings, and research visits, the establishment of nuclear fuel reserve for an emergency, the supply of nuclear material and equipment, controlled thermonuclear fusion, including multilateral projects, and others. The implementation of projects in the territory of both countries would be carried through their individual laws, regulations, and treaties for the peaceful use of nuclear energy. This article also upholds that, although both states are in agreement for peaceful usage of nuclear energy, it would not affect the "unsafeguarded nuclear activities of either Party" [2]. Therefore, under this Agreement, there will be no interference of both states in activities associated with military usage of nuclear technology.

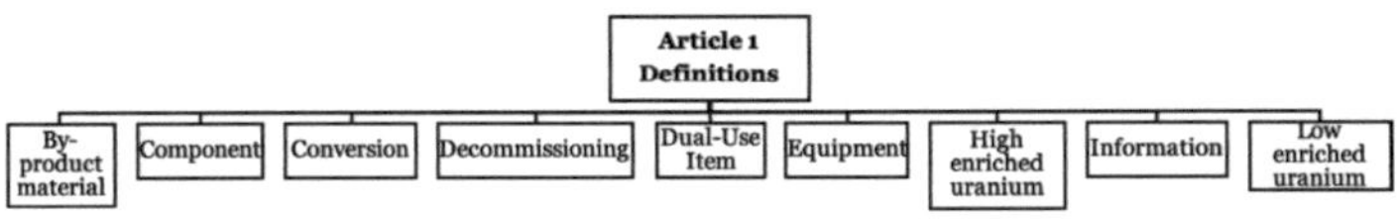

Figure 3.
Article 1 Definitions 1 [2].

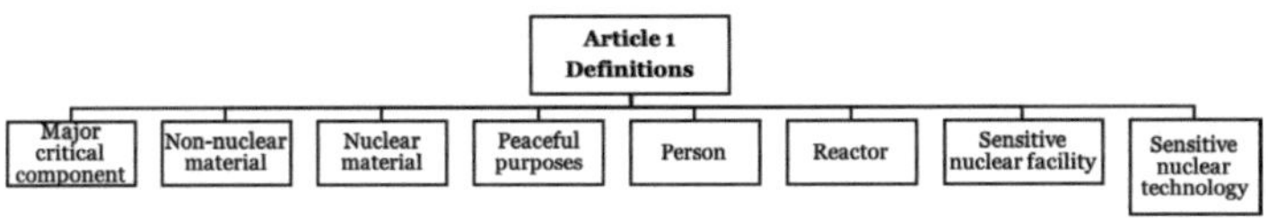

Figure 4.
Article 1 Definitions 2 [2].

Articulating details about the transfer of nuclear energy *via* reports, computer programs, and data banks, Article 3 [2] covers the following areas about research and development of nuclear reactors, highly developed research in nuclear science and technology, nuclear safety and security, and environmental concerns, fuel cycle undertakings to meet civil nuclear energy requirements, consideration of nuclear power roles in national energy plans to use nuclear energy for healthcare, agriculture, and industry. There will be no transfer of "Restricted Data" [2], which is not under the provisions of this agreement.

Article 4 [2] enables nuclear trade between both states on subjects of their mutual interest and wherever applicable trade with a third party as well. Furthermore, for the uninterrupted functioning of nuclear facilities, the industry in both states must reassure reliable supplies to nuclear reactors. The authorization procedure for export-import licenses for bilateral trade and nuclear material movement should be sound and efficient.

Concerning the transfer of nuclear and nonnuclear material, equipment, and technology, Article 5 [2] provides the following instructions:

- Nuclear components would be transferred under the regulatory framework of this agreement. Any fissionable materials transferred under this agreement must be low-enriched uranium, which shall be transferred as nuclear fuel to reactors.

- Sensitive critical nuclear components, such as technology, material, and nuclear facilities, can be used for reprocessing and enrichment by laws.

- If both parties agree, a limited amount of "special fissionable material" can also be transferred to use as detectors, standards, and samples.

Under Article 6 [2] both countries can carry out the following "Nuclear Fuel Cycle Activities":

- Under the "territorial jurisdiction" use of uranium (U) and its 20% enriched U-235 can be carried out.

- High-enriched uranium (HEU), U-233, plutonium (Pu), and other irradiated nuclear material can be transferred, used, or produced through nuclear and non-nuclear material in accordance with this deal.

- As mentioned in the Joint Statement released in 2005 about civil nuclear cooperation between both states, they can grant consent to each other about the reprocessing of nuclear material, and for this purpose, India would establish a new national reprocessing facility under IAEA safeguards. All arrangements under this Article should include provisions about—Environmental concerns (Article 11), Physical Protection (Article 8), and Storage Standards (Article 7).

Article 7 [2] states that all sensitive nuclear material transferred or produced under this deal shall always be stored at the facilities with the IAEA level of protection as explained in document INFCIRC 225/REV 4. Both states may contain details about the facilities of each other for further activities. This article also prohibits both states from transferring any of the sensitive nuclear material, equipment, or technology to any unauthorized person or party without mutual consent.

Physical protection of sensitive nuclear content shall be of utmost priority for both states. For this purpose, Article 8 [2] demand from both states to follow IAEA document INFCIRC/225/Rev.4 as titled, "The Physical Protection of Nuclear Material and Nuclear Facilities" [2]. It would be the responsibility of both states to keep each other updated through diplomatic sources ensuring the physical protection of nuclear material in their "territorial jurisdiction" [2]. Under the provision of this Article, there would not be any interference of India and the USA in each other's peaceful nuclear activities. Subsequently, Article 9 [2] stipulates that nuclear material used, transferred, or produced in accordance with this deal would not be used for any military purpose of making any nuclear explosive device.

Article 10 [2] sets the provisions for safeguards of nuclear content to be maintained by both states. India in pursuant with its safeguards agreement with IAEA agreed to follow norms set up by IAEA at its territory to safeguard the nuclear material, equipment, and technology under this agreement. Similarly, the U.S. territory safeguards of nuclear content will be in accordance with the agreement signed between the USA and IAEA for the application of safeguards in the USA on

November, 18, 1977 at Vienna. At the non-application of safeguards, and measures as decided by IAEA, it may check and confer both states for verification measures. As stated in Article 11 [2], both states must check the environmental impact of radioactive, chemical/thermal contamination and work together to minimize the impact of these activities on the environment and physical health.

To put this deal into effect, Article 12 [2] commands both states to conduct nuclear activities under their territorial jurisdiction on time with in-depth management practices. Both states should avoid interference in nuclear activities being held at the other's territory; also, they must not interfere in the nuclear policy of the other parties. As per Article 13 [2], both countries shall consult each other for the successful implementation and development of nuclear projects and their actions also shall be in congruence with the objectives of Article 2. A "Joint Technical Working Group" [2] would be set up to check the Administrative Arrangement.

About the termination and cessation of this deal, Article 14 [2] provides the right to both states to terminate this deal with a prior one-year written notice explaining the reason for termination. However, before terminating this agreement, both countries may consult each other to discuss the reasonability of termination as stated in Article 13 [2]. If a party that is seeking termination refers to a violation of the agreement as a reason for termination, it shall consider whether the action was caused unconsciously or purposely breaching the Vienna Convention on the Law of Treaties. If it cites a violation of IAEA safeguards as a reason for termination, then compliance with the IAEA board of directors would be a crucial factor to count on. If the agreement is terminated then both states must return acquired nuclear material, equipment, technology, and other components to each other. Article 15 [2] guides both states to resolve their disputes through negotiations and consultations.

Article 16 [2] holds that this agreement will be effective for 40 years and shall be extended for the next 10 years, except if there is a notice for agreement termination from either party. If the agreement is ceased to exist due to termination or expiration, Articles 5.6 (c), 6, 7, 8, 9, 10, and 15 would not be affected if nuclear components remain in "territorial jurisdiction" of the concerned parties. Lastly, Article 17 [2] asks to set up an "Administrative Agreement" [2] to facilitate the efficient implementation of this cooperation.

2.3 India-France civil nuclear cooperation

A cooperation agreement for the peaceful usage of nuclear energy was signed between the Government of India and the Government of France in 2008.

Article 1 [3] of this agreement asks both countries to use nuclear energy for nonmilitary purposes, according to the obligation of international law and individual nuclear policy of each state. This article is divided into three subparts further discussing areas of cooperation, forms of cooperation, and nuclear safety.

To fulfill the obligations for cooperation, Article 1 [3], and Article 2 [3] discuss that the Memoranda of Understanding (MoU) signed by the designated authorities must include detailed scientific and technical programs and planning about scientific and technical exchanges. Transfer of nuclear material, equipment, and setting up of nuclear facilities under this deal would be directly between the parties, and such a transaction to any third party must be in persuasion from applicable authority. Toward the implementation of the agreement and individual domestic laws, Article 3 [3] states that all provisions under this agreement must agree to their respective administrative, tax, and customs measures. Article 4 [3] talks about the cooperation between both India and France in building, and commissioning of nuclear power plants under the regulatory framework. This Article also promotes cooperation among involved operators to develop equally suitable settings for mutual collaboration.

About the supply and accessibility of the nuclear power plant, Article 5 [3] holds that the country under whose jurisdiction the nuclear power plant is positioned would provide uninterrupted and reliable access to reactor systems, nuclear fuel supplies, and other mechanisms for the lifetime of the given nuclear power plant. In this regard, "supply of nuclear fuel for the lifetime of India's safeguarded reactors" [3] a longstanding deal according to Article 2 [3] would be established between entitled authorities from India and France. Therefore, to prevent the disorder of supply, it would be the responsibility of France to support India to acquire a strategic reserve of nuclear fuel *via* summoning friendly countries or joining a group that can arrange or restore fuel supply to India during the supply disruption. The recycling and modification of nuclear content transferred, used, and produced under this agreement would be conducted at the national nuclear facilities under the IAEA safeguards in the recipient country.

In the interest of the particular industries, utilities, and consumers, Article 6 [3] says that both India and France must recognize the reliability of nuclear supplies as a crucial need for both countries and it must be on time. There must be "progressive localization and indigenization" [3] for the effective operation of nuclear installations wherever applicable. Under Article 7 [3], there must be the protection of intellectual property rights by the responsible designated authorities in both countries. Both India and France must set up a required regulatory framework as referred to in MoU in Article 2 [3], for the protection of nuclear material and technology produced and transferred under this agreement.

The entitled authority from both countries would look after liability issues such as "Civil Nuclear Liability" under Article 8 [3]. According to the recognized international doctrines, both India and France must form a "Civil Nuclear Liability Regime" [3] to compensate for any damage instigated by nuclear material, technology, equipment, and facilities. For this purpose, Article 9 [3] holds that all nuclear components used, produced, and transferred under this agreement would be used only for nonexplosive and peaceful purposes.

Article 10 [3] under the provisions of Article 5 [3], guides both states to maintain IAEA safeguards for nuclear materials, technology, and equipment used or transferred under this agreement, in accordance with their individual safeguards' agreement with IAEA. The safeguards of the nuclear material transferred to the French Republic under this nuclear cooperation would be according to the agreement signed in 1978 and revised in 1998 between France, the European Atomic Energy Community (EURATOM), and IAEA. Similarly, an agreement signed between the Government of India and IAEA would look after the safeguarding of nuclear material on Indian territory. Moreover, if in the future IAEA notes that IAEA safeguards are not being maintained, then it will require to consult with both parties for the verification.

As mentioned in Article 11 [3], the nuclear material obtained or recovered, technology, equipment, and facilities under this agreement would be "transferred or retransferred beyond the jurisdiction of recipient party" under Article 15 [3]. Also, the termination of safeguards, when nuclear material is fully consumed or no longer in use-would be decided by the pact between the Government of India and France, EURATOM, IAEA, and other involved agencies.

About the confidentiality and usage of the shared information, Article 12 [3] says that information and technical data used under the agreement would be considered sensitive and discretionary and must not be shared with any third party without each other's consent, also both countries must ensure security and safety of the shared information. Article 13 [3] asks both parties to make sure that all nuclear components, as mentioned in Article 11 [3] would be maintained exclusively by the designated authority of the person in charge.

To ensure the physical protection of the nuclear components Article 14 [3] emphasizes the effective implementation of national legislation and international conventions, such as IAEA's Convention on Physical Protection of Nuclear Material (CPPNM), which was signed in 1979, entered into force in 1987 and amended in 2005.

Article 15 [3] describes the condition that needs to be fulfilled by both countries if they wish to transfer nuclear components as discussed in Article 11 [3] to any third party. Such transfers and retransfers can be only conducted after obtaining written assurance from a third party under IAEA safeguards about the peaceful and nonexplosive usage of nuclear components. India and France first provide written consent to each other before transferring nuclear material to any third party except the member states of the European Union. Article 16 [3] holds that this agreement between India and France would not affect the other international obligations of each of them for the peaceful use of nuclear energy.

Promoting an equal partnership between both countries, Article 17 [3] articulates the relationship of both states to be reliable and stable to discuss the implementation of the agreement. A Joint Committee needs to be established and this committee must meet from time to time to discuss the further implementation of the agreement. Article 18 [3] prohibits amendments in the agreement throughout the period of the agreement. Both parties shall first provide mutual written consent for the proceeding of the amendment procedure. If any amendment is planned to be introduced in the agreement, it shall be first ratified and approved by both states under their individual statutory framework. Article 19 [3] forms Annexes of this agreement to be an essential part of this agreement.

Discussing the entry into force, duration, and termination of the agreement, Article 20 [3] says that after completing their internal process both states shall report each other, which will lead to the "entry into force" of the agreement. The duration of the agreement is 40 years and would be renewed automatically for another 20 years if not notified by each of the parties is to terminate the agreement. If either of the party demands termination of the agreement before the expiry of the agreement it shall provide 12 months prior notice with the reason for termination to the concerned party. If the agreement is not renewed the specific provisions under Article 2 [3] would persist to be applicable and if the agreement is terminated Articles 7, 8, 9, 10, 11, 12, 13, 14, 15, and 16 [3] would continue to be applied wherever applicable.

2.4 India-Russia civil nuclear cooperation

Desiring to further promote the relationship of strategic partnership by developing cooperation between the two states, India and Russia agreed to sign a civil nuclear cooperation agreement toward the peaceful applications of nuclear energy.

Following the principles of sovereignty, equality, and noninterference, Article 1 [4] addresses both states to develop bilateral cooperation for peaceful nuclear use under the regulatory framework of their individual legislations, international commitments, and national nuclear programs. This article further comprehends areas of cooperation between both states as follows [4]:

- Applied and advanced research for use of nuclear energy for healthcare, environment, industry, agriculture, and climate change.

- Development, maintenance, and safety of nuclear energy infrastructures, such as nuclear reactors, power plants, and radiation sources.

- Industrial production of isotopes and other components for use in nuclear reactors.

- Also, exploration and development of thorium and uranium deposits for peaceful purposes.

Above mentioned area of cooperation must be carried out through—joint programs and working groups for the implementation of specific projects, conferences, workshops, and symposia for mutual consultations, training, and exchange of scientific personnel in addition to relevant information and equipment. The fourth section of this Article articulates the objective of this agreement as a step toward "peaceful nuclear cooperation" [4] and both states would not interfere in any of the nuclear activity of each other which does not come under the regulatory framework of this agreement.

Article 2 [4] explains the important terminology used in this agreement as follows (**Figures 5** and **6**):

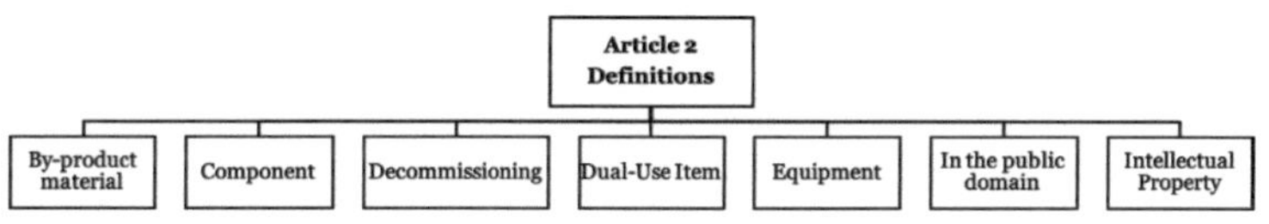

Figure 5.
Article 2 Definitions 1 [4].

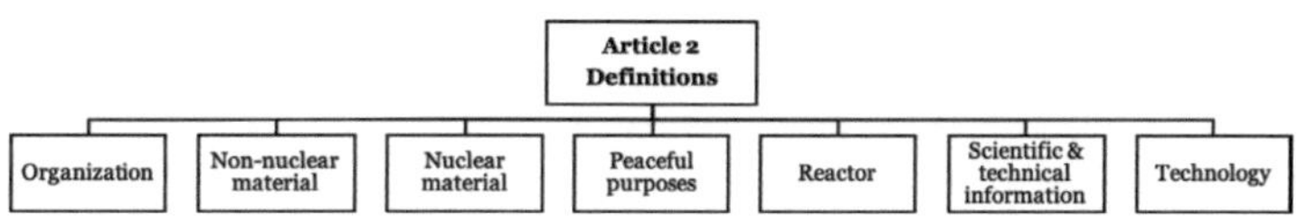

Figure 6.
Article 2 Definitions 2 [4].

Various sections of Article 3 [4] comprehend steps for the unhindered implementation of the agreement as follows:

- Designated authority to implement contracts would be—the "Department of Atomic Energy" of the Government of the Republic of India [4], and the State Atomic Energy Corporation "Rosatom" for the Government of the Russian Federation [4].

- Under the obligations of this agreement, "Department of Atomic Energy" and "Rosatom" would execute and conclude programs and projects between India and Russia. The scope of cooperation, rights, financial, and other legal terms and conditions shall be in congruence with individual regulations.

- It will be the responsibility of both designated authorities to handle nuclear and nonnuclear material, equipment, information and technology, and facilities used and transferred under this agreement with safety and security.

Article 4 [4] demands from both states to consider information and documents exchanged and transferred under this agreement to be "confidential" [4] in accordance with individual legislation in both countries. In the case of Russia, it shall be adopted as an "official information of limited distribution" [4] and under regulations of the Indian government, it would be treated as "confidential" [4] with access only to regulated authorities and personnel. The right to intellectual property is discussed in Article 5 [4]. Only designated organizations under the jurisdiction of this agreement from both countries are allowed to transfer intellectual scientific information to each other. The rights of the jointly created or produced scientific information would belong solely to India and Russia and if either of them wishes to transfer this information to any third party, they must first acquire written consent from each other to do so.

About nuclear enrichment, Article 6 [4] states that this cooperation allows Uranium-235 enrichment up to the level of 20% [4]. The enrichment and reprocessing of nuclear fuel should be performed under IAEA safeguards in concurrence with the "Agreement between the Government of India and the IAEA for the application of Safeguard to Civilian Nuclear Facilities, document-INFCIRC/754" [4] in the Indian case. Correspondingly, under Russian jurisdiction, "Agreement between the USSR and the IAEA for the application of safeguards in the Union of Soviet Socialist Republic (USSR), document-INFCIRC/327" [4]. This article also instructs both states to outline a separate agreement for, "transfer of technology and facilities for chemical reprocessing of irradiated fuel, isotopic uranium enrichment and heavy water production" [4].

The nuclear components received, transferred, and/or produced under this deal shall not be used for military or explosive purposes, as mentioned in Article 7 [4]. Both states must consult the IAEA document, "Physical Protection of Nuclear Material and Nuclear Facilities document-INFCIRC/255/Rev 4" [4], for the protection measures. Both countries must use nuclear technology only for "declared purposes" and must not transfer any of the technology, material, or information to any third party.

As per Article 8 [4], any damage caused by a nuclear accident, and its liability, shall be concluded between India and Russia according to their international commitments and respective legislations. Under Article 9 [4], both countries shall establish a Joint Coordination Committee to facilitate faster and smoother implementation of this agreement. The committee would constitute "competent authorities" [4] from both states, which must consult each other over issues that may arise due to any dispute. Article 10 [4] holds that to amend this agreement, both parties must acquire written consent from each other. Article 11 [4] mentions the entry into force process of the agreement and the duration of the agreement, which is 40 years, and its further extension procedure for the next 10 years after the completion of 40 years of cooperation.

2.5 India-UK civil nuclear cooperation

With the intention to strengthen bilateral strategic partnership through cooperation in the peaceful use of nuclear energy, the Government of India and the Government of the United Kingdom of Great Britain and Northern Ireland signed an Agreement on 13 November 2015 in London, and on 16 December 2016, this agreement was entered into force [5].

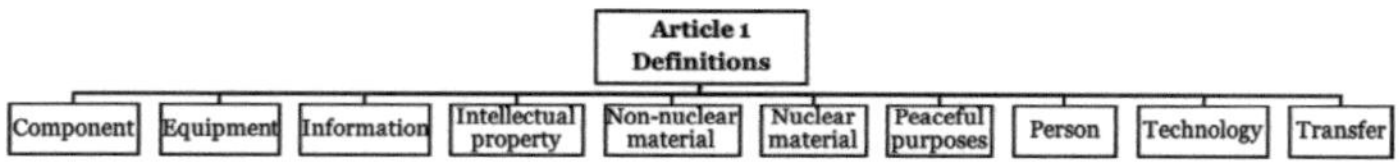

Figure 7.
Article 1 Definitions [5].

Article 1 [5] of this agreement defines important terminology as used in legal documents (**Figure 7**).

To foster peaceful usage of nuclear energy, Article 2 [5] elaborates following areas of cooperation between both states [5]:

- Through research and development, make use of nuclear energy to generate electricity and water desalination and others in the field such as agriculture, healthcare, industry, and medicine.

- Environmental concerns and keep a check on nuclear safety and radiation aspects.

- Develop a strategic reserve of nuclear fuel to manage the nuclear fuel cycle if the supply gets disrupted in any case.

- Technical training and assistance associated with nuclear technology and science.

This Article holds that provisions under this agreement would not affect other individual aspects associated with nuclear energy as well as individual international commitments and national policies and legislations.

Article 3 [5] suggests various forms of cooperation to fulfill objectives explained in Article 2 [5] through the supply of nuclear and nonnuclear material, equipment and technology, training and exchange of personnel, organization of workshop and seminars for participation by scientific and technical staff and others. Nuclear trade of nuclear material, technology, equipment, and other components can be carried out between both states and associated authorities as per Article 4 [5].

Transfer of all material whether between India and UK or any authorized third party would be conducted only when decided by both states in a written statement under Article 5 [5]. This article also states that there would not be any transfer of nuclear components outside the territorial jurisdiction of each party. To retransfer obtained nuclear material to any third party, firstly recipient party shall obtain written consent from the transferring party and a letter of commitment to use nuclear only for the peaceful purpose under the IAEA safeguards from a third party. However, these requirements would not be applicable to the Member States of the European Union (EU) under the EURATOM Treaty [5].

In pursuant to Article 6 [5], processing, reprocessing, and storage of nuclear material under this agreement should be carried out only at sites that have been reported by India to the IAEA as per document INFCIRC/754 [5]. Also, this article allows a maximum of 20% enrichment of U-235 [5], which will be assigned to India under agreement provisions of INFCIRC/754 [5]. Article 7 [5] commands both states to make sure that nuclear components and products would be used only for peaceful purposes and handled carefully under IAEA safeguards. Safeguards of nuclear material and other components transferred to the UK would be subject to regulations of Chapter 7 of EURATOM Treaty and IAEA document INFCIRC/263, including an Additional Protocol (AP) of IAEA, which was signed in 1998 and entered into force in 2004 [5]. And, nuclear material transferred to India would be subject to IAEA safeguards under the guidance of IAEA document INFCRIC/754 along with an Additional Protocol, which was signed in 2009 and entered into force in 2014 [5]. Therefore, if IAEA agrees upon any possibility of non-application of its safeguards, then both parties need to discuss and adopt applicable verification measures in written format.

Article 8 [5] outlines the applicable scope of regulations of the agreement. All nuclear material, equipment, and technology transferred, used, or produced under this agreement would not be subject to the regulations of this agreement if both states decide through a written statement that they are not subject to obligations of the agreement or they have been assigned or reassigned above mentioned nuclear components to the third party beyond their jurisdiction. Article 9 [5] demands adequate actions from both parties to ensure the protection of nuclear material and technology, these protection measure obligations shall be carried out as per the recommendations of IAEA document INFCIRC/225/Rev.5 [5] titled as, "Nuclear Security Recommendation on Physical Protection of Nuclear Material and Nuclear Facilities" [5].

To protect intellectual property used, transferred, or produced under the provision of this deal. Article 10 [5] asks India and UK to prevent the unauthorized use of nuclear material, technology, and information hereby, this article under this agreement forbids the transfer of intellectual property rights to any unauthorized agency or third. Article 11 [5] to expedite the effective implementation of this nuclear cooperation, ask to provide the administrative arrangement from appropriate government authorities to implement the required obligation. Similarly, under Article 12 [5] it would be the responsibility of both states to consult each other from time to time for the effective execution of projects under this agreement. If there arises any refute, both countries shall seek negotiations in accordance with the founding document of MoU to settle issues between them as per Article 13 [5].

Article 14 [5] allows both states to amend this agreement from time to time if necessary. About the entry and duration, Article 15 [5] notes that this agreement would be in force when both states notify each other about the fulfillment of required internal legalities in written. The duration of this agreement is 40 years, which is subject to extension for the next 20 years if neither of the countries shows intentions to terminate it at least 6 months prior to the agreement expiry. If notified to be terminated by either party, Article 16 [5] asks both parties to cooperate in the termination procedure. Both parties shall provide one-year prior notice with a reason for each other to terminate this agreement. If the agreement is terminated in any case than the obligations of Articles 5, 6, 7, 8, 9, and 10 of this agreement would not be affected and shall continue to be in force.

3. Japanese case analysis

In this section, authors have reviewed and analyzed the Japanese civil nuclear cooperation and agreements with IAEA, USA, EURATOM, Russia, and UK.

3.1 Japan-IAEA civil nuclear cooperation

An agreement between Japan and IAEA was signed in 1977 to fulfill the obligations of Articles III.1 and 4 of the Non-Proliferation Treaty (NPT) [6]. Therefore, in Article 1 government of Japan upholds support to Article III.1 of NPT and accepts IAEA safeguards on all special fissionable material from all sources use up only for peaceful nuclear activities under its jurisdiction inside and outside Japan. Article 2 [6] and 3 [6] covers the procedures to apply safeguards provided by IAEA in pursuant to the given agreement, hereby the Japanese government must maintain an accounting system (National Systems of Safeguards) to bring all nuclear and associated components in regulation. For the effective implementation of the safeguards, the IAEA taking into consideration the findings of the National System would conduct autonomous measurements and observations to check if nuclear material is not used for nuclear weapons and explosive devices. Article 4 [6] asks both parties to assist each other in the effective implementation of the agreement.

Articles 5 [6], 6 [6] and 7 [6] provide a framework to implement safeguards in a manner that they must not intervene and impede the nuclear and technological developments of Japan. Both parties must safeguard confidential and sensitive data subject to the provision of this agreement and must not share such information with any third party except with the Board of Governors of the Agency. Article 8 [6] sets the provisions for Japan to share information regarding nuclear material and related activities subject to safeguard with IAEA. In accordance with Article 9 [6], IAEA must appoint an Inspector to carry out activities efficiently under the given agreement in consensus with the Japanese government. This Agency Inspector would check if the confidentiality of sensitive data is being maintained by both parties, and inspection of nuclear activities in Japan's territory is not hampering its nuclear activities. Relevant privileges and immunities must be provided by the government of Japan to the IAEA Inspector and other officials as per Article 10 [6]. Article 11 [6] holds that IAEA safeguards are subject to termination if IAEA decides that nuclear material is no longer usable for any nuclear activity. As per Article 12 [6], if the Japanese government requires transferring its nuclear material to any other country, then it must notify the Agency about the nuclear transfer and its safeguards, and Agency must maintain documentation of each transfer and retransfer of nuclear material. Article 13 [6] mentions that both parties must agree to use nuclear material for nonnuclear activities before its termination from the agreement due to dilution.

In accordance with Article 14 [6], if the government of Japan decides not to apply the given safeguards to maintain its discretion, it must inform the IAEA board and shall affirm that nuclear material would be used only for peaceful nuclear activities and during the period of "non-application of safeguards" [6] nuclear material very strictly would not be producing explosive devices and nuclear weapons. Article 15 [6] explains about the finance and Article 16 [6] mentions the liability of the third-party in the case of nuclear damage. Any dispute between both parties regarding the agreement implementation must be resolved in concordance with international law as per Article 17 [6]. Articles 18 [6] and 19 [6] provide "measures in relation to verification of non-diversion" of nuclear activities. Article 20 [6], 21 [6] and 22 [6] discusses the interpretation of the Agreement and that both parties must consult each other for the interpretation and implementation of the agreement as well as hold negotiations to resolve any dispute. Safeguards enforced by IAEA upon Japan under other agreements would be suspended on the entry of this agreement according to Article 23 [6]. As per the provisions of Article 24 [6], there will be the same conditions like "entry into force" [6] to amend this agreement for which both parties must consult each other. About the duration and "entry into force," Article 25 [6] says that when the Japanese government would notify IAEA Director-General about the completion of its constitutional and statutory requirements the agreement would "enter into force" and would be in force till Japan is a participant in NPT. Article 26 [6] forms the attached protocols as an integral part of this agreement.

Objectives for safeguards are highlighted from Article 27 [6] to 30 [6], the foremost aim would be to keep a check on the usage of nuclear material and ensure that they must not be used to manufacture nuclear weapons and devices. For this purpose, "material accountancy" has been suggested as a significant safeguard measure that would help to contain the illicit usage of nuclear material. Therefore, Article 31 [6] and 32 [6] emphasize the role of the National System of Safeguards as an authority in Japan that may facilitate verification activities for nuclear maintenance and inspection in accordance with Article 3 [6]. Articles 33 [6] and 34 [6] set up areas for safeguards application as well as restricting their application to material in mining ore or ore processing activities. Article 35 [6], pursuant to Article 13 [6] states that on mutual consensus between Japan and IAEA, IAEA safeguards would be terminated given the nuclear material is extremely exploited and irrecoverable. Articles 36 [6] and 37 [6] provide an exemption to special fissionable material, plutonium-238 (exceeding 80%), and nuclear material (used in nonnuclear activity as per Article 13) from safeguards application at the request of the Japanese government. However, Article 38 [6] states that if exempted nuclear material is processed or stored with nuclear material under safeguards, then safeguard provision must be reapply on such material. Article 39 [6] and 40 [6] commands both parties to make the required "Subsidiary Arrangement" to fulfill duties effectively under the given agreement and must be enforced within 90 days from the date of entry into force. After the release of the first report, it will be the responsibility of the IAEA to create an inventory of all nuclear material in Japan liable to the given agreement as per Article 41 [6].

Provisions and information regarding the design of nuclear facilities are articulated in Article 42 [6] to 45 [6]. The "design information" regarding all nuclear facilities subject to this agreement must include information, such as—facility identification, description of general arrangement at the facility, layout of equipment used, report of prevailing and propositioned processes regarding nuclear material accounting and surveillance, measurement of flow, and inventory undertakings. Article 46 [6] enlist the objectives to regulate material balance, get detailed nuclear material verification, to create record and reports of evolution procedure, and others regarding the above-mentioned "design information". Articles 47 [6] and 48 [6] set forth a clause of reexamination in case of changes in nuclear operations and verification by the Agency Inspector regarding the design information as provided in Articles 42 [6] to 45 [6]. Articles 49 [6] and 50 [6] command Japan to provide information to IAEA with respect to nuclear material when used outside authorized facilities. Provisions for the record system are set out from Articles 51 [6] to 58 [6]. Designated under "Subsidiary Arrangements" the Japanese government must maintain a record system of all nuclear material equipment, operating facilities, and associated components to be retained at least every 5 years.

Articles 59 [6] to 61 [6] articulate general provisions to prepare reports based on records as mentioned in Articles 51 [6] to 58 [6]. Articles 59 [6] to 61 [6] explain the general provisions of the reports system. An "Accounting Report" and "Special Report" must be prepared in English, Russian, French, or Spanish but otherwise as specified in the "Subsidiary Arrangement" with respect to nuclear material subject to IAEA safeguards. A detailed description of the "Accounting Report" is mentioned in Articles 62 [6] to 67 [6]. First of all, an initial report of all nuclear material must be released by the Japanese government to IAEA within a month of "entry into force." Each material balance area report must cover inventory changes and a material balance statement centered on the physical inventory of nuclear material located in the material balance area. Twice a year, IAEA shall provide a statement book consisting of inventory change entries of all nuclear materials subject to safeguards of this agreement. A "Special Report" pursuant to Article 68 [6] must be prepared by the government of Japan in case of any unusual contamination and loss of nuclear material exceeding the specified limit. The government of Japan must provide an explanation and clarification of any report if requested by IAEA as per Article 69 [6].

Obligations with respect to inspection by Agency are provided in Articles 70 [6] to 73 [6]. The purpose of such inspection would be to authenticate initial report information and identify changes as well as check the composition and quantity of nuclear material at the time of its transfer in and out of Japan later Articles 74 [6] and 75 [6] discusses the scope of the inspection to be conducted in accordance with Articles 51 [6] to 58 [6]. Entry to Agency Inspector shall be provided for inspection of all locations mentioned in the initial report with respect to nuclear material, as per Articles 76 and 77 [6]. Articles 78 [6] to 82 [6] provide guidelines about the frequency and intensity of inspection and hence states that IAEA must conduct a yearly inspection of authorizing nuclear facilities, material balance areas, and other components in consultation with the Japanese government. It will be the responsibility of the IAEA to provide the Japanese Government prior to notice concerning the visit of the Agency Inspector to their nuclear facilities pursuant to Articles 83 [6] and 84 [6]. As mentioned in Articles 85 [6] and 86 [6], officials nominated by the IAEA and accepted by the government of Japan shall be appointed by the Director General of Agency within a month of the agreement's "entry into force." Articles 87 [6], 88 [6] and 89 [6] emphasize the conduct of the Agency Inspector during the visit to carry out activities carefully without hampering the construction of facilities and nuclear operation activities. Article 90 [6] commands IAEA to release statements mentioning the result of verification activities conducted during the inspection.

To fulfill the objectives of this agreement, pursuant to Article 91 [6] Japan must ensure the safeguard of nuclear material to be transferred in an out of Japan. For this purpose, Articles 92 [6] to 96 [6] provide guidelines for the export and import of nuclear material in Japan. To export nuclear material out of Japan, the Japanese government must notify the IAEA board about the quantity of nuclear material, safeguard measures, state to which it would be transferred, and date of nuclear material dispatch and arrival, also both parties must agree upon the implemented procedures. A similar process will be applicable there will be a transfer of nuclear material into Japan. Pursuant to Article 68 [6], Article 97 [6] commands the Japanese government to prepare a Special Report in case of any delay or loss of nuclear material during international transfer.

Article 98 [6] provides definitions of the important terminology for the effective interpretation and implementation of this agreement (**Figures 8** and **9**).

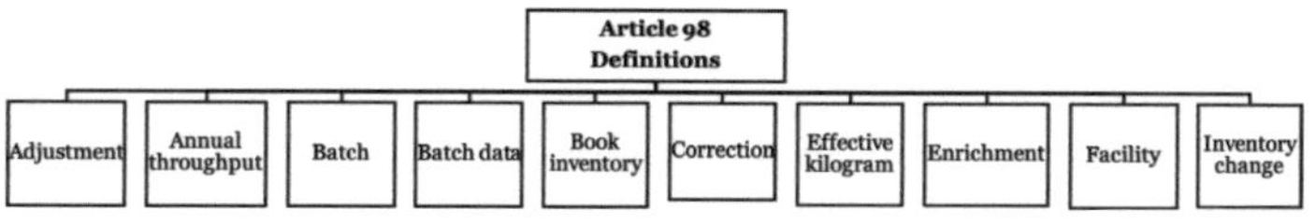

Figure 8.
Article 98 Definitions 1 [6].

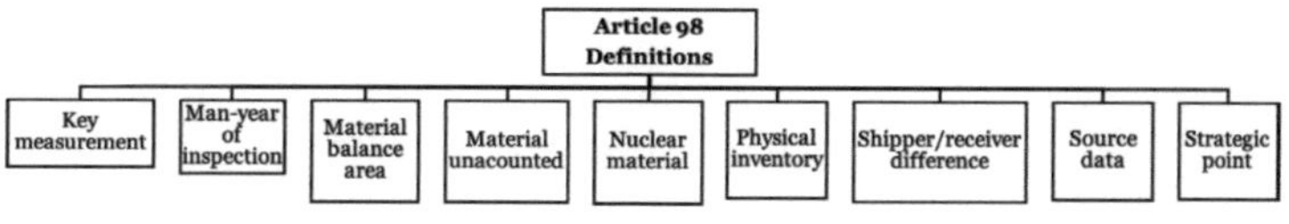

Figure 9.
Article 98 Definitions 2 [6].

3.2 Japan-USA civil nuclear cooperation

Japan and the US have been the longest-standing partners promoting the peaceful use of Civil Nuclear Energy, their first nuclear agreement was signed in 1955 and then in 1968. In 1987, Japanese government and the US government revised their "previous agreement" [7] and signed a new agreement reasserting their commitment to ensuring the research and development of nuclear energy for peaceful purposes and support objectives of IAEA toward the international observance of the Non-Proliferation Treaty. The articles of this agreement are discussed in detail as follows:

For the purpose of this agreement, definitions of the below-mentioned terminologies are explained in Article 1 (**Figure 10**) [7]. Article 2 [7] provides the following guidelines for both states to cooperate for the effective implementation of the agreement:

- On the agreed terms, there shall be an exchange of information on the matter of healthcare, nuclear safety, environmental considerations, and others between both states.

- Designated person or authority from both parties can share nuclear material, equipment, and nuclear technology under the provisions of this agreement. However, restricted data and sensitive nuclear technology specified by both states would not be transferred under this agreement

- Authorized officials from both countries can render services at each other's territory on the matters subject to this agreement.

- The usage of nuclear material and all nuclear activities either in the territory of Japan or under its jurisdiction anywhere else would be considered a step in meeting Japan and IAEA's agreement to support the Non-Proliferation Treaty. Also, in the case of the United States, all of its nuclear activities at its territory or under its jurisdiction would be carried as a fulfilling step toward the application of safeguards as per the US-IAEA agreement.

All nuclear material, technology, equipment, and associated components used and shared between both states would be subject to the provisions of this agreement. Above mentioned nuclear components would cease to be part of the agreement if they are transferred beyond the dominion of the receiving party or the nuclear material is consumed or diluted to the extent that it is no longer usable for any nuclear activity.

About the storage of sensitive nuclear material, such as Uranium-233, Plutonium, and high enriched uranium. Article 3 [7] commands both countries on mutual consensus to store such material at the facilities upholding IAEA safeguards. Article 4 [7] asks to transfer sensitive nuclear components and special fissionable material only to a person authorized by the receiving party. If agreed mutually both parties can

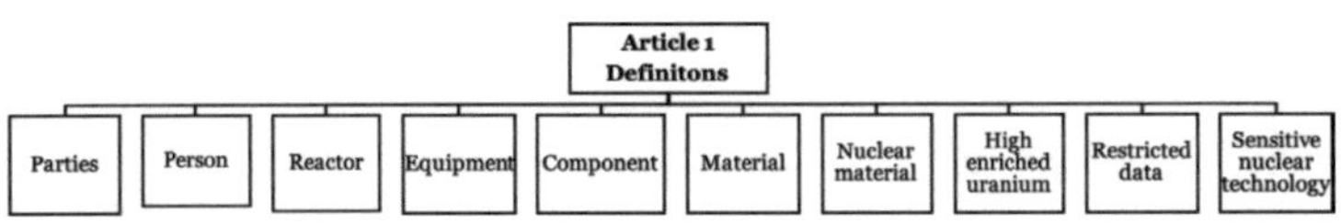

Figure 10.
Article 1 Definitions [7].

reprocess and alter the form of nuclear material transferred, used, or produced under this agreement as per Article 5 [7]. Article 6 [7] limits the enrichment of the U-235 to 20%. Article 7 [7] demands suitable physical protection of all the nuclear components subject to the provision of this agreement. Therefore, this agreement prohibits both states from using nuclear components for any military purposes or developing nuclear explosive devices as per Article 8 [7].

Article 9 [7] is about the safeguard of nuclear components in accordance with IAEA safeguards. Nuclear material and associated components transferred, used, and produced under this agreement in the territory Japan and the US would be safeguarded in pursuant to their respective safeguard agreement with IAEA. With respect to nuclear material, technology, equipment, and components used under this agreement, Article 10 [7] set forth rights equivalent to Articles 3 [7], 4 [7], 5 [7], 6 [7], and 12 [7] of this agreement to other third parties or state, which is in agreement with either of both parties. With respect to their national security, Article 11 [7] obliges both countries to make consistent efforts to prevent nuclear proliferation and make special arrangements to facilitate Articles 3 [7], 4 [7], and 5 [7] of this agreement to foster peaceful use of nuclear energy. About the termination of this agreement, Article 12 holds that if either Japan or the US fails to comply with the provisions of Articles 3 [7], 4 [7], 5 [7], 6 [7], 7 [7], 8 [7], 9 [7], and 11 [7] of this agreement, defies IAEA provided safeguards or does not follow the decision of arbitral tribunal both states have right to terminate this agreement at any given time and demand the return of any material, nuclear material, and equipment transferred under this agreement.

On the date of the entry into force of this agreement, Article 13 [7] would terminate the previous agreement. However, projects initiated under the previous agreement would be resumed under the provisions of this agreement. To promote bilateral cooperation pursuant to this agreement, Article 14 [7] asks both states to consult with each other through diplomatic channels and other forums to resolve the issues regarding the interpretation and implementation of this agreement. However, if any such issue is not resolved through consultation, mediation, or negotiations both parties on mutual consensus must submit issues to an arbitral tribunal and the decision of the tribunal would be obligatory for both countries. Article 15 [7] forms Annexes of this agreement as to its integral part, which can be modified on mutual consensus without any amendment. Article 16 [7] holds that upon completion of their respective legal provision, both states must notify each other, and hence agreement must enter into force within a month from the notification. The duration of this agreement is 30 years, which may be continued thereafter until either of both parties seeks termination. Nevertheless, before the completion of 30 years, either party may seek termination by providing 6 months prior to the other party. Articles 1 [7] to 9 [7], 11 [7], 12 [7], and 14 [7] of this agreement would continue to be in effect on the applicable matters even after the agreement is terminated. Lastly, on the mutual consensus, both countries may amend or change this agreement with a new agreement.

3.3 Japan-EURATOM civil nuclear cooperation

Article 1 [8] defines terminologies from the legal document for effective interpretation of agreement as follows (**Figure 11**). To promote peaceful nuclear trade and research activities between Japan and the European Union "Scope of Cooperation" through the establishment of a research and development institute, the nuclear fuel cycle industry is discussed in Article 2 [8]. This Article further elaborates on forms through which this cooperation can be fostered between both parties in the following manner:

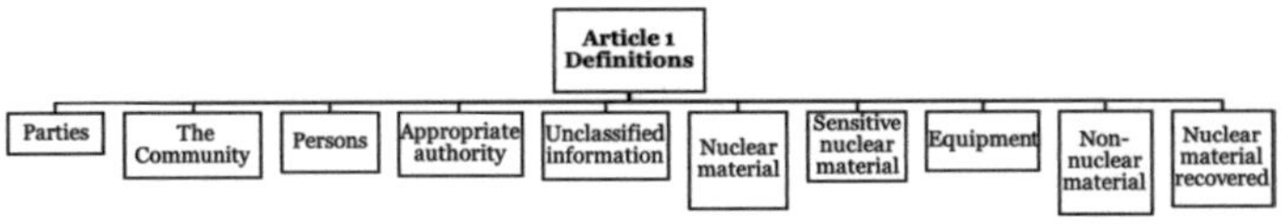

Figure 11.
Article 1 Definitions [8].

- On the agreed terms both parties may share nuclear material, technology, and equipment as well as supply and exchange unclassified information with each other.
- In pursuant to the regulatory framework of the agreement, authorized persons from both parties can perform and receive the nuclear fuel cycle and other services.
- To nurture bilateral cooperation in this direction, there shall be an exchange of experts between both parties.
- They must encourage cooperation between each other through consultations and negotiations.

Article 3 [8] is about the nuclear and associated materials under the territorial jurisdiction of Japan, it holds that all nuclear and nonnuclear material and equipment, whether shared between Japan and Agency or transferred from any other third party to Japan would be subject to the provisions of this agreement. However, above mentioned nuclear material and equipment would not be subject to this agreement if they would be transferred beyond the territorial jurisdiction of the recipient party or if they are no longer usable for any nuclear activity and mutually decided by both parties that they are no longer subject to this agreement. Article 4 [8] further elaborates on the cooperation as discussed in Article 2 [8] on nuclear research and development. For this purpose, both Japan and the EU shall encourage the involvement of researchers and organizations from universities, laboratories, and private sectors.

To facilitate the effective implementation of the agreement, Article 5 [8] asks both parties to implement provisions of the agreement in good faith to be coherent with "prudent management practices required for the economic and safe conduct of their nuclear activities" [8]. The provision of this agreement must not be used for interference in the nuclear policy of either party as well as not for any industrial or commercial advantages. Regarding the mixing procedure of nuclear material, it should be handled on the "principle of fungibility and proportionality" during the process of enrichment, fuel fabrication, and conversion. In pursuant to international rules and regulations, Article 6 [8] asks both parties to safeguard the protection of intellectual property transferred used and produced under this agreement. Hereby, Article 7 [8] commands both parties to use nuclear material and equipment subject to this agreement only for peaceful purposes.

Article 8 [8] requires the application of safeguards pursuant to the EURATOM Treaty and IAEA agreement from Japan and the EU community. Article 9 [8] prohibits both parties from retransferring nuclear beyond their jurisdiction to any third party to do so they must obtain a prior consent letter from each other and an assurance letter of not using nuclear material for any military purpose from the country to which nuclear material would be retransferred. There shall be transparency between both parties with respect to the management of nuclear and nonnuclear material, equipment, and technology in accordance with Article 10 [8] of the given agreement. Article 11 [8] commands both states to ensure the physical protection of nuclear material and associated components transferred, used, or produced during this agreement. During the International transfer of nuclear material in confirmation with Japan and European members of the Community, they must follow the provision of the "Convention on the Physical Protection of Nuclear Material."

Pursuant to Article 12 [8], provisions of the given agreement shall be observed corresponding to the "Agreement between the Government of Japan and the Government of the United Kingdom of Great Britain and Northern Ireland for Co-operation in the Peaceful Uses of Nuclear Energy" and the "Agreement between Government of Japan and Government of French Republic." The obligations of the above-mentioned bilateral agreement beyond this agreement must continue to be in effect. Article 13 [8] articulates suspension and termination clauses of the present agreement and states that, if the Japanese government or any Member state of the Community violates the decision of the Arbitral Tribunal or provisions of Articles 7 [8], 8 [8], 9 [8], and 11 [8] of this agreement both parties have right to terminate cooperation under the given agreement may demand the return of the nuclear material transferred pursuant to this agreement. Article 14 [8] asks both parties to designate applicable authorities for the effective implementation of "operational procedures."

Both parties must consult each other in case of any difference regarding agreement implementation and must opt for negotiations and conciliation; however, if the dispute is still not resolved they shall submit the dispute to Arbitral Tribunal under Article 15 [8]. The decision undertaken by the Arbitral Tribunal would be obligatory for both parties. Article 16 [8] creates Annexes as an integral part of this agreement. Article 17 [8] fixes the duration of this agreement for 30 years from the date of "entry into force" with 55 years of automatic extension if either of both parties does not seek termination of the agreement before its expiry. Articles 7 [8], 8 [8], 9 [8], and 11 [8] of the given agreement would remain effective upon the agreement termination.

3.4 Japan-Russia civil nuclear cooperation

Japan and Russia signed a bilateral civil nuclear cooperation agreement for the peaceful uses of nuclear energy on May, 12, 2009 in Tokyo. To fulfill the objectives of this agreement Article 1 [9] provides a definition of the following mentioned terminologies (**Figure 12**).

Under the regulatory framework of this agreement and their respective legislations, Article 2 [9] provides guidelines for Japan and Russia to cooperate with each other in the following ways:

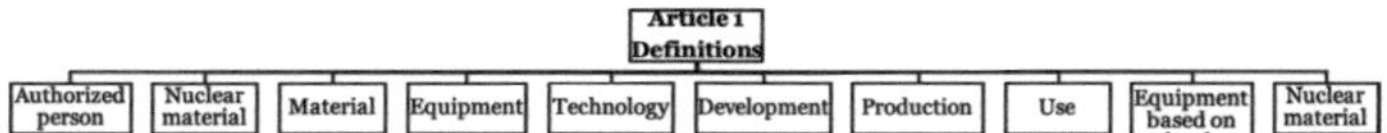

Figure 12.
Article 1 Definitions [9].

- Through the exchange of researchers and specialists, the exchange of information is not only limited to nuclear technology but also nuclear safety and other associated aspects.

- Transfer of nuclear material, equipment, and technology on agreeable terms between both states.

- There shall be the provision of services and receipt of that services between both states whenever required by either of the Contracting Party.

Both countries must assist each other to construct and operate light water reactors, for the application of radio-isotopes and radiation, exploration and utilization of Uranium, processing of radioactive waste, and most importantly nuclear safety from radiation. However, this Article also clarifies that technology and nuclear material that is not covered in this agreement would not be shared by either country.

Article 3 [9] states that to fulfill the above-mentioned cooperation, both states need to apply safeguards as directed by IAEA. If nuclear material is transferred to Japan, then it shall apply safeguards as per "the Safeguard Agreement" [9] signed between the Government of Japan and IAEA to implement the "Non-Proliferation Treaty" in 1977. And if the nuclear material recipient in Russia, it must oblige "the Safeguard Agreement for Russian Federation" signed with IAEA in 1985. Article 4 [9] obligates both states to use nuclear material, equipment, and technology used, transferred, and produced under this agreement only for peaceful purposes for the welfare of society. Therefore, to fulfill above mention obligations, Article 5 [9] binds all sensitive nuclear material and technology subject to provisions of the "Safeguards Agreement" of Japan and Russia and must be placed at facilities as selected by IAEA. Similarly, Article 6 [9] asks both states to observe provisions of "Convention on Early Notification 1986" [9], "Convention on Assistance in the Case of a Nuclear Accident 1986" [9], "Convention on Nuclear Safety 1994" [9], and the "Joint Convention on the Safety of Spent Fuel & Radioactive Waste Management 1997" [9] for the effective implementation of projects under this agreement.

To ensure the physical protection of transferred nuclear material, equipment, and technology Article 7 [9] asks both parties to adopt provisions of the "Convention on the Physical Protection of Nuclear Material 1980" and "International Convention for the Suppression of Acts of Nuclear Terrorism 2005." As per Article 8 [9] any of the associated nuclear components must not be transferred or retransferred to any unauthorized third party without the prior written consent of each other. Article 9 [9] limits the enrichment of nuclear material, such as U-235, to 20%. All nuclear material and associated components shared between Japan and Russia would automatically fall under the provisions of this agreement once it is entered into force pursuant to Article 10 [9]. Article 11 [9] sets provision for both states to not use this agreement as a base for interfering in each other's industrial or commercial interest or for any other purpose, which may hinder the peaceful usage of nuclear energy. This article also states that nuclear items when used for enrichment, mixed processing, fuel fabrication, conversion, and reprocessing must be operated under the "principle of fungibility and proportionality" [9].

If nuclear material is transported to a state beyond the jurisdiction of the recipient party Article 12 [9] holds that in such a situation it will no longer be answerable to the provision of this agreement. Also, if both states mutually decide from the perspectives of "nuclear safeguards" about particular nuclear material that it is no longer usable for any nuclear activity it would not be subject to provisions of this agreement. Article 13

[9] holds that secret and classified information of both states must not be exchanged in the name of this agreement.

Article 14 [9] asks both states to consult each other in case of any disagreement and conduct negotiations, conciliation, and mediation to resolve the dispute. However, if the dispute is still not resolved either of both states may appeal Arbitral Tribunal to settle the matter. Following all legal arbitral procedures, the decision of the tribunal would be fixed and obligatory in both states. Article 15 [9] provides cases that may lead to the suspension or termination of this agreement as follows:

- If either of the party fails to observe the provisions of Articles 4 [9], 5 [9], 6 [9], 7 [9], 8 [9], and 9 [9] of the agreement or disregard the decision of Arbitral Tribunal as mentioned in Article 15 [9].

- If safeguards provided by IAEA are defied or terminated by either party.

- If the nuclear explosive device is detonated by Russia using nuclear items transferred from Japan under this agreement, then Japan has the right to seek termination of the agreement, and similarly if Japan detonates a nuclear explosive device, the Russian government can seek termination of the agreement.

If either of the party decides to terminate this agreement, they may first consult to take corrective measures. This article will be applicable only in case when either of the party fails to adopt corrective measures in a given time frame.

In accordance with applicable international commitments, safeguards, and respective legislation, Article 16 [9] asks both states to ensure the protection of intellectual property created, used, and/or transferred under this agreement. Article 17 [9] observes Annexes of this agreement as an integral part of this agreement, it also states that this agreement can be amended through mutual understanding of both states provided through a written contract. In the last, Article 18 [9] fixes the duration of this agreement to 25 years, which may be further extended thereafter until terminated. To seek termination, either of both parties must provide 6 months prior notice to each other before the expiry of the agreement. However, agreement termination would not affect Articles 1 [9], 4 [9], 5 [9], 6 [9], 7 [9], 8 [9], 9 [9], 12 [9], 14 [9], and 15 [9] as these Articles would continue to be in effect even after termination.

3.5 Japan-UK civil nuclear cooperation

To foster the peaceful use of nuclear energy, Article 1 [10] commands both parties (Japan and UK) to cooperate with each other in the following ways:

- There shall be an exchange of experts between both states from their respective organizations to boost organizational cooperation between them.

- In accord with mutual terms and conditions, they must share nuclear material, equipment, and “unclassified information” [10] between contracting parties or person of each other’s jurisdiction.

- On the matters subject to this agreement, authorized “party or person” [10] would be allowed to obtain assistance as well as perform at each other’s territory.

Under Article 2 [10] cooperation between both states shall be in accordance with each other's law and regulations, regarding the safeguard of nuclear facilities and nuclear material. In the case of Japan, regulations would be subject to "Agreement between the Government of Japan and the Agency in Implementation of Article III. 1 and 4 of the Non-Proliferation Treaty" [10], done on March 1977. Similarly, "Agreement between the United Kingdom of Great Britain and Northern Ireland, EURATOM and the Agency in connection with the Non-Proliferation Treaty, done in September 1976" [10] must be contemplated for the application of safeguards provided by IAEA. Therefore, nuclear material, equipment, technology, and other components transferred, used, or produced under this agreement would be used only for peaceful purposes under Article 3 [10].

To fulfill the above-mentioned obligations, Article 4 [10] says that the safety of nuclear material in Japan's territory would be pursuant to the Japan-IAEA agreement, while on the territory of the United Kingdom along with its agreement with IAEA, safeguards applied by EURATOM as per the 1957 Treaty will also be applicable. And if any of the clauses of the EURATOM treaty is not applicable to the provisions of this agreement, both parties must consult to adopt rectifying measures corresponding to agreement commitments. Article 5 [10] demands adequate "physical protection" of the nuclear material used, transferred, or produced during the application of the agreement.

Nuclear material and other components subject to the obligations of this agreement would not be transferred to any unauthorized party under Article 6 [10]. However, if there arise any condition to transfer nuclear components, then both states must obtain prior written consent from each other, also a written commitment from the third party to use nuclear components for peaceful purpose only. Additionally, products such as equipment for reprocessing and enrichment, 20% or more enriched U-233, U-235, or plutonium would not be reassigned by the recipient country to any third party.

Article 7 [10] would bring nuclear material and related components of both parties under the regulatory framework of the agreement except in the following situations:

- If nuclear components have been transferred outside the dominion of the recipient Contracting Party as per this agreement.
- Also, if both states mutually decide about particular items not subject to agreement.
- If IAEA decides that nuclear material is no longer exploitable for any nuclear activity or is practically irrecoverable.

Discussing termination of the previous agreement between both countries. Article 8 [10] holds that, once this agreement would be in force it shall terminate the previous agreement. Also all nuclear material and related components would fall under the obligations of this agreement. There shall not be any prejudice based on obligations imposed on either party at the signing of the agreement as per Article 9 [10].

Article 10 [10] requires both parties to consult each other if there arises any issue regarding the implementation of the agreement. If any dispute or issue cannot be resolved through consultation or negotiation, therefore at the appeal of either Contracting Party or issue must be submitted to Arbitral Tribunal. If within 30 days of the appeal of the Arbitration Tribunal, either Contracting Party has not nominated an arbitrator then either Party may request the International Court of Justice to employ an arbitrator, and a decision made by Tribunal would be obligatory on the Contracting Party. As per Article 11 [10] if either of both parties does not comply with the provisions of the agreement, then they have the right to ask the defiance party to adopt cooperative measures and if these corrective measures are not adopted by the party which is not complying with provisions of the Agreement, then other party has right to terminate or suspend this agreement.

Article 12 [10] provides a description of the mentioned terms for the purpose of this agreement (**Figure 13**). The Annexes provided with this agreement are an essential part of this agreement and can be revised on the mutual consensus of both countries as mentioned in Article 13 [10]. Article 14 [10] provides guidelines about entry, duration, and termination procedure of this agreement. The agreement would be enforced for the next 25 years and thereafter, upon getting written notification from both states about the completion of their respective legislative procedures. To terminate the given agreement both parties shall provide 6 months prior notice to each other either on the completion of the initial 25 years or any time later on. Articles 3 [10], 4 [10], 5 [10], 6 [10], 7 [10], 10 [10], and 12 [10] of this agreement would not be affected by the agreement termination and must continue to be in effect. Therefore, on February, 251,998, Government of Japan and the Government of the United Kingdom of Great Britain and Northern Ireland signed this agreement to form strategic cooperation toward the peaceful use of nuclear energy [10].

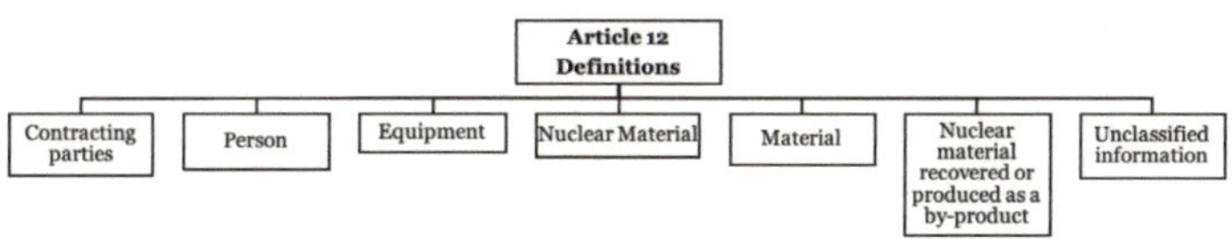

Figure 13.
Article 12 Definitions [10].

4. India-Japan civil nuclear cooperation agreement

This section explains the background of the India-Japan civil nuclear cooperation, present status, and further detailed analysis of the India-Japan bilateral civilian nuclear cooperation.

4.1 Background and present status

The background and present status of India-Japan bilateral civilian nuclear cooperation is described in **Figure 14**. As shown in **Figure 14**, in 2010, India and Japan

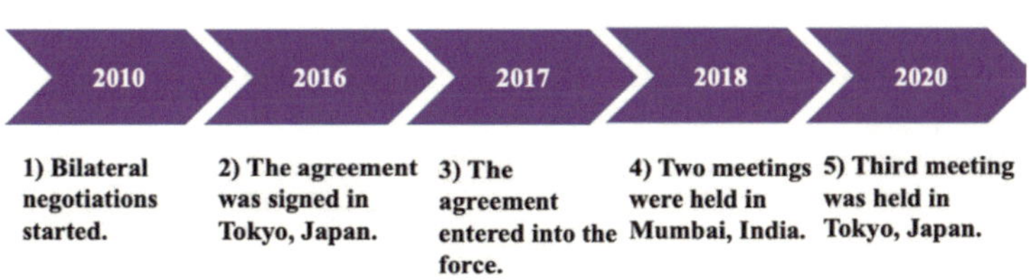

Figure 14.
Timeline of India-Japan civil nuclear cooperation [11–16].

started negotiations for bilateral civilian nuclear cooperation. After 6 years-long negotiations, in 2016, the agreement was signed in Tokyo. The agreement was entered into force in 2017. In the direction of working group meetings, two meetings were held in Mumbai, India, in 2018, and in 2020, the third meeting was held in Tokyo, Japan.

4.2 Analysis

To fulfill the commitments of this agreement, Article 1 [12] defines the meaning of important terms in the context of a legal agreement document between the Government of the Republic of India and the Government of Japan in order to avoid vagueness and facilitate effective interpretation of the agreement (**Figure 15**). Fostering bilateral cooperation between India and Japan for peaceful and nonexplosive usage of nuclear energy. Article 2 [12] emphasizes that both parties shall implement international laws and applicable nuclear treaties in furthering this deal. This Article further provides ways through which cooperation can be nurtured between both parties as follows:

- *Via* exchange of scientific experts and information between the designated authority or person from both countries. It will be the responsibility of both states to facilitate the entry and stay of scientists and experts in their respective territories.

- Make available nuclear material, technology, and equipment to each other to carry on projects pursuant to this agreement effectively.

- Facilitating services for activities such as designing, construction, and operation of nuclear reactors and others by a party and receipt of services by other parties under the regulatory framework of this agreement.

- Both parties must cooperate with each other on all aspects of the nuclear fuel cycle, nuclear safety and security, and application of radio-isotopes and radiation in healthcare, agriculture, industry, and environmental protection.

Article 3 [12] binds both parties to use nuclear material, technology, equipment, and by-products transferred, used, or produced under this agreement only for peaceful and nonexplosive purposes. Article 4 [12] instructs both India and Japan to apply protective safeguards measures pursuant to the IAEA safeguard agreement with India and Japan, respectively. Also, appropriate verification measures are required from both parties if the Agency determines that the safeguards application is unachievable. Next Article 5 [12] is asking both countries to keep an account of all nuclear material transferred, produced as well as recovered pursuant to this deal. Also, there will be

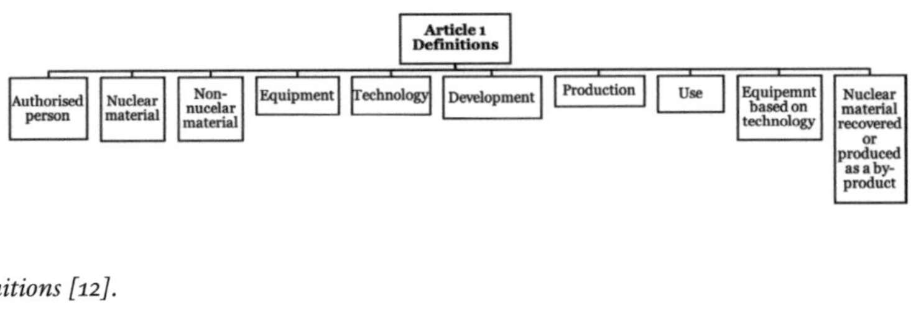

Figure 15.
Article 1 Definitions [12].

an exchange of nuclear information and technology between both parties under the provision of this agreement. Both India and Japan are required to reiterate their commitment toward International Conventions concerning nuclear security and safety to which parties and members as per Article 6 [12].

To maintain the physical protection of all nuclear material and associated components Article 7 [12] commands both parties to adopt adequate protection measures in accordance with the regulations of applicable international conventions such as "Convention on the Physical Protection of Nuclear Material & Nuclear Facilities." It will be the individual responsibility of both states to ensure the physical protection of nuclear material under their territorial jurisdiction as guided by the IAEA document INFCIRC/225/Rev. 4. In pursuant to international agreements consulting intellectual property rights, Article 8 [12] asks both parties to protect the intellectual property, information, and technology acquired during this agreement from any unauthorized use and prohibits any disclosure of such information to any third party. Article 9 [12] encourages nuclear trade between the designated authority or person from both countries, however, the provisions of the agreement must not be misused for obtaining commercial gains. Article 10 [12] strictly prohibits the transfer or retransfer of nuclear material to any third party beyond the jurisdiction and to do so written consent must be obtained from the supplying party.

About nuclear enrichment, Article 11 [12] restricts the enrichment of uranium-235 isotope to less than 20% transferred and used in accordance with this agreement. It also states that nuclear material recovered as a by-product will be reprocessed under the jurisdiction of the Republic of India complying with the Agreement signed in 2009 between India and IAEA for the "Application of Safeguards to Civilian Nuclear Facilities." Article 12 [12] covers the subject matter related to the transfer of nuclear material, equipment, and technology between both states. Both parties must sign the Memoranda of Understanding to exchange scientific and technical information regarding the above-mentioned components between designated persons from both parties. Once the nuclear material and associated components are shared between both states, thereby upon entering into their respective territory, they will fall under the respective regulatory framework of both states. It is further stated that if nuclear material is transferred beyond the territorial jurisdiction of the state of the receiving party or is exploited in a manner that no longer usable for any nuclear activity, it shall no longer be subject to the given agreement. To cultivate proper handling of nuclear material, "principles of fungibility, proportionality, and equivalence" to nuclear material transferred, used, and produced under this agreement.

To promote the effective implementation of the agreement, Article 13 [12] asks both states to establish Joint Committee and Technical Working Group to view matters regarding the implementation of projects under this agreement. If in case there is any dispute or disagreement regarding the implementation of the agreement between both states, they must opt for negotiations and consultations to resolve their dispute. If either of both parties violates agreement provisions, Article 14 [12] provides the right to both countries to terminate the agreement before its expiry by providing 1-year prior notice to each other. If the agreement is terminated both states also have the right to demand the return of nuclear material and equipment transferred under this agreement from each other as applicable under law.

As per the provisions of Article 15 [12], this agreement would not influence the individual rights and international commitments of both parties with respect to nuclear and nonnuclear material, technology, and equipment not subject to the given agreement. Article 16 [12] is about the amendment procedure and Annexes of this agreement; it holds that a given agreement can be amended at any given time if provided with written agreement from both parties; it also binds Annexes as an integral part of the agreement which can be modified without amendment. Article 17 [12] fixes the duration of the given agreement to 40 years from the date of entry. The agreement can be extended for the next 10 years automatically if either of the parties does not submit a written document of termination of the agreement no later than 6 months before its expiration. Upon the termination of agreement provisions of Articles 1 [12], 3 [12], to 7 [12], 10 [12], 11 [12], 12 (paragraph 3) [12], 13 [12], and 14 [12] would not be affected and would continue to be in effect.

5. Summary and conclusion

This chapter provides an overview of Indian and Japanese civilian nuclear cooperation with other countries as well as India-Japan mutual cooperation on the basis of Government documents for the peaceful uses of nuclear energy. To summarize and conclude the chapter, the authors highlight the importance of international civil nuclear cooperation [17] toward the smooth implementation of bilateral as well as multilateral civilian nuclear deals. The Government' documents guide both agreed parties and make sure the timely implementation of the signed agreements with accuracy and assurance. Importantly, due to the complexity of nuclear power in terms of safety, security, and safeguards regimes, authors recommend establishing an efficient operation of bilateral and multilateral civilian nuclear cooperation to ensure the peaceful uses of nuclear energy as well as minimize the nuclear proliferation risk. To enhance the transparency in peaceful nuclear development in countries or regions, mutual cooperation through civilian nuclear agreements between or among the Governments plays a vital role. In brief, this chapter facilitates to understand Indian and Japanese nuclear relations through the lens of various described civilian nuclear agreements.

Author details

Saurabh Sharma* and Masako Ikegami
Tokyo Institute of Technology, Tokyo, Japan

*Address all correspondence to: sharma.s.ab@m.titech.ac.jp

References

[1] IAEA. Agreement between the Government of India and the International Atomic Energy Agency for the Application of Safeguards to Civilian Nuclear Facilities, INFCIRC/754. 2009

[2] DAE, Government of India. Important Agreements-Agreement between the Government of India and the USA on Cooperation in Peaceful Uses of Nuclear Energy, 10 October 2008. 2008

[3] DAE, Government of India. Important Agreements-Agreement between the Government of India and the Government of France on Cooperation in Peaceful Uses of Nuclear Energy, 30 September 2008. 2008

[4] DAE, Government of India. Important Agreements-Agreement between the Government of the Republic of India and the Government of the Russian Federation on Cooperation in the Use of Atomic Energy for Peaceful Purposes, 12 March 2010. 2010

[5] DAE, Government of India. Important Agreements-Agreement between the Government of India and the Government of United Kingdom of Great Britain and Northern Ireland for Co-operation in the Peaceful Uses of Nuclear Energy, 13 November 2015. 2015

[6] IAEA. The Text of the Agreement of 4 March 1977 between Japan and the International Atomic Energy Agency in Implementation of Article III.1 and 4 of the Treaty on the Non-proliferation of Nuclear Weapons, INFCIRC/255, March 1978. 1978

[7] U.S. Government Printing Office Washington. Proposed Agreement Between the United States and Japan Concerning Peaceful Uses of Nuclear Energy. In: 100th Congress, 1st Session, House Document 100-128, Committee on Foreign Affairs, 09 November 1987. 1987

[8] Ministry of Foreign Affairs (MOFA) of Japan. Agreement between the Government of Japan and the European Atomic Energy Community for Co-operation in the Peaceful Uses of Nuclear Energy, 2006. 2006

[9] MOFA. Agreement between the Government of Japan and the Government of the Russian Federation for Cooperation in the Peaceful Uses of Nuclear Energy, 2009. 2009

[10] Agreement between the Government of the United Kingdom of Great Britain and Northern Ireland and the Government of Japan for Co-operation in the Peaceful Uses of Nuclear Energy. Treaty Series No. 36. 1999 [Previously Published as Japan No. 1 (1998) Cm 3948]

[11] Sharma S, Ikegami M. Research review on India-Japan civil nuclear cooperation agreement towards the progress of peaceful uses of the nuclear energy–原子力平和利用進展にむけた日印原子力協定のレビュ–. In: Abstract Proceedings and Poster Short Presentation at the 39th Annual Meeting of Institute of Nuclear Materials Management (INMM). Japan: The University of Tokyo; 2018

[12] Ministry of Foreign Affairs of Japan. Agreement between the Government of Japan and the Government of the Republic of India for Cooperation in the Peaceful Uses of Nuclear Energy. 2016. Available from: https://www.mofa.go.jp/mofaj/files/000202920.pdf

[13] Library of Congress. Japan/India: Diet Approves Civil Nuclear Cooperation Agreement. 2017. Available from: http://

www.loc.gov/law/foreign-news/article/japanindia-diet-approves-civil-nuclear-cooperation-agreement/

[14] Ministry of Foreign Affairs of Japan. Meetings on Japan-India Civil Nuclear Cooperation. 2018. Available from: https://www.mofa.go.jp/press/release/press4e_001973.html

[15] Ministry of Foreign Affairs of Japan. Meetings on Japan-India Civil Nuclear Cooperation. 2018. Available from: https://www.mofa.go.jp/press/release/press1e_000089.html

[16] Ministry of Foreign Affairs of Japan. Meetings on Japan-India Civil Nuclear Cooperation. 2020. Available from: https://www.mofa.go.jp/press/release/press4e_002521.html

[17] Singh A, Sharma S, Kalra MS. Levelised cost of electricity for nuclear power using light water reactor technology in India, Economic & Political Weekly. 2018;**2018**:10

Section 2

Fundamentals

Chapter 3

Influence of Prompt Neutron Emission on the Final Distribution of Mass, Kinetic Energy, and Charge of Fragments from Actinide Fission

Modesto Montoya

Abstract

In nuclear fission of actinides samples, the fragments emit prompt neutrons before they reach the detectors. This chapter shows the effects of that emission on the final distribution of mass, charge, kinetic energy, and prompt neutron multiplicity of fragments. Those effects depend on the experimental technique. This chapter shows the curve of the maximal value of total kinetic energy as a function of primary fragment mass for the reactions $^{233}U(n_{th}, f)$, $^{235}U(n_{th}, f)$, and $^{239}Pu(n_{th}, f)$ which reflect the deformation properties of fragments in their ground states and the Coulomb effect on their scission configurations. The odd–even effects on the cold fission region will be also presented.

Keywords: thermal neutron fission, mass yield, kinetic energy, charge, prompt neutron multiplicity

1. Introduction

In 1939, Otto Hahn and Fritz Strassman discovered the fission of the uranium nucleus induced by neutrons [1], which was experimentally corroborated by Meitner and Frisch [2]. In 1939, von Halban et al. [3] discovered that fission fragments emit neutrons. To interpret the experimental results, Bohr and Wheeler [4] proposed a model in which the potential surface of the fissile nucleus is a function of several variables that represent the deformation of the nucleus, which starts from a well of stability to rise through the potential surface and reach the saddle point. From the saddle point, the system descends through a fission valley (where the attractive nuclear interaction and the repulsive Coulomb interaction compete) to the scission point where finally two separate complementary fragments are formed. The interpretation of the fission process may be based on the potential surface of the fissile system [5].

From the scission point, the two independent fragments are repelled by electrostatic interaction until they reach their final kinetic energy. On the way, the fragments decay mainly by gamma and neutron emission.

In that scheme, to investigate the nature of fission, it results necessary to know at least the characteristics of the final state of the fission process, i.e., the scission configuration. For this reason, the goal of most fission experiments is to see how the process might manifest itself in the distribution of primary fragment mass and kinetic energy distribution.

In this chapter, the effects of the emission of prompt neutrons on the distribution of mass, charge, and kinetic energy of the fragments, compared to the primary quantities will be reported. We will see how the experimental technique affects the results of their measurements.

2. Kinetic energy distribution as a function fragment mass

Let us list some definitions and relations to be used to show the effects of neutron emission on the distribution of mass, charge, kinetic energy, and average prompt neutron multiplicity of the fission fragments.

The fissile nucleus is characterized by its charge and mass Z_0 and A_0, respectively. The primary complementary fragments are characterized by their masses (A, A'), charges (Z, Z'), and kinetic energies (E, E'), respectively.

The conservation laws of mass and linear momentum are expressed in the relations.

$$A_0 = A + A', \tag{1}$$

and

$$AE = A'E', \tag{2}$$

respectively.

The total kinetic energy is.

$$TKE = E + E', \tag{3}$$

The final complementary fragments masses.

$$m = A - n \tag{4}$$

and

$$m' = A' - n', \tag{5}$$

where n and n' are their corresponding number of emitted prompt neutrons.

For each isobaric fragmentation, the values of n and n' are given approximately by the relations.

$$n = \overline{\nu}\left(1 - \frac{TKE - \overline{TKE}}{\alpha} + r\right) \tag{6}$$

and

$$n' = \overline{\nu}'\left(1 - \frac{TKE - \overline{TKE}}{\alpha} - r\right), \tag{7}$$

where $\overline{TKE}$ is the average total kinetic energy, $\overline{\nu}$ and $\overline{\nu}'$ are the average prompt neutron multiplicities, respectively, α is a parameter taken from experimental data, and r is a correlation term.

Due to the emission of neutrons, the values of the final kinetic energy of the fragments will be, approximately,

$$e = E\left(1 - \frac{n}{A}\right), \tag{8}$$

and

$$e' = E'\left(1 - \frac{n'}{A'}\right) \tag{9}$$

These relations are formulated considering no recoil effect by neutron emission. If one simulates an isotropic prompt neutron emission relative to each fragment center of masses, the result is a kinetic distribution characterized by an average obeying approximately the same relations with standard deviations depending on the emitted neutron kinetic energy. The result is equivalent to a lower resolution in kinetic energy measuring.

The total kinetic energy of the final fragments is defined as

$$tke = e + e' \tag{10}$$

Experimentally one has access to the values of final fragment mass and kinetic energy. The fragment masses measured by the double energy method (2E), defined as provisional masses of the complementary fragments, μ and μ', are calculated using the conservations relations.

$$A_0 = \mu + \mu', \tag{11}$$

and

$$\mu e = \mu' e'. \tag{12}$$

From the Eqs. (8), (9), (11), and (12), one obtains.

$$\mu = A_0 \frac{e'}{e' + e} = A_0 \frac{E'}{E' + E \times F} = A_0 \frac{A}{A + A' \times F}, \tag{13}$$

where

$$F = \frac{1 - n/A}{1 - n'/A'}. \tag{14}$$

The provisional mass μ may be lower or higher than A, depending on F. For the reactions ^{233}U(n$_{th}$, f), ^{235}U(n$_{th}$, f), and ^{239}Pu(n$_{th}$, f), there is a fragment mass region where $\overline{\nu} \gg \overline{\nu}'$. In this case $\mu > A$. Obviously, m will be equal or lower than A.

2.1 Kinetic energy distribution as a function of the final fragment mass

In 1978, using the Monte Carlo method, the experiment to measure the distribution of e as a function of m of fragments from the fission induced by thermal neutrons of ^{235}U was simulated. The simulation was carried out before the corresponding experiment. The simulated standard deviation of the final light fragment kinetic energy σ_e showed a peak of 9 MeV for $m = 109$ in a mass region where the primary standard deviation is approximately 5 MeV. This peak was the result of the neutron emission combined with the effects of the slopes of the mass yield and the average kinetic energy of the fragments. Experimental results, obtained on the LOHENGRIN mass spectrometer, confirmed that prediction [6]. See **Figure 1**.

In 1983, Behafaf et al. repeated that experiment extending the domain of measured masses to the heavy region, finding a peak in the standard deviation for $m = 125$, which was not reproduced by their Monte Carlo simulation [7]. See **Figure 2**. In 2007, through another Monte Carlo simulation, Montoya et al. reproduced the mentioned peak [8]. See **Figure 3**. The origin of the discrepancy between those results is the discontinuity between $m = 124$ and $m = 125$ on the curve of $\overline{e}$ that Montoya et al. used as input for the simulation while probably was not used in the Belhafaf et al. simulation.

2.2 Kinetic energy distribution as a function of provisional fragment mass

In 1978, using the 2E technique, Asghar et al. measured the quantities $\overline{e}$, σ_e, $\overline{tke}$, and σ_{tke} as a function of provisional mass μ of fragments from the reactions ^{235}U(n$_{th}$, f) and

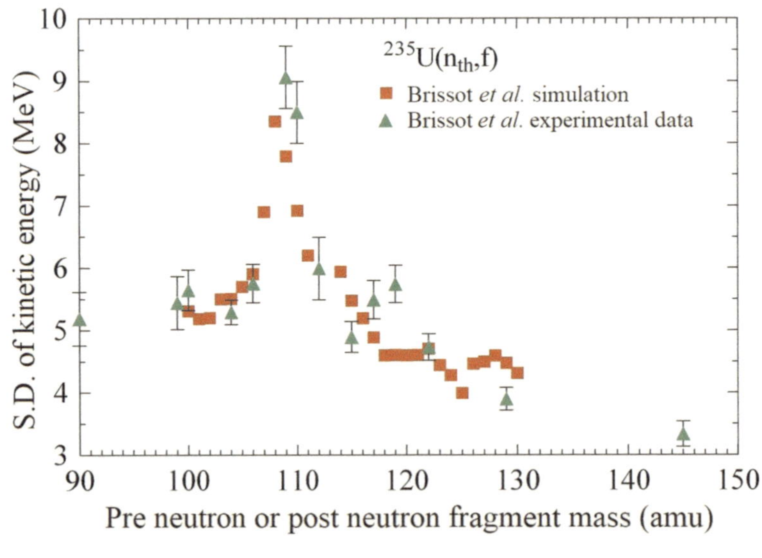

Figure 1.
Standard deviation of the final kinetic energy distribution as a function of the final mass (m) of fragments from fission of ^{235}U induced by thermal neutrons. The output of the Monte Carlo simulation (squares) is to be compared with the experimental result (triangles) [6].

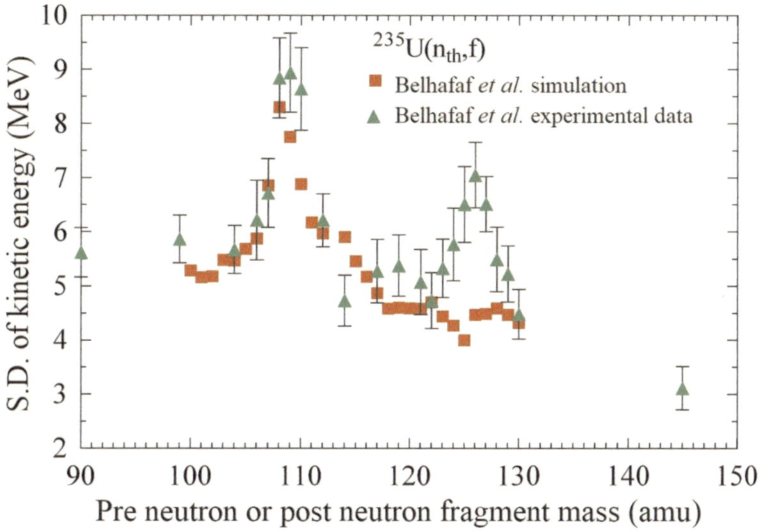

Figure 2.
Standard deviation of the final kinetic energy distribution as a function of the final mass (m) of fragments from fission of ^{235}U induced by thermal neutrons. The output of Monte Carlo simulation (squares) does not reproduce the experimental peak (triangles) at m =125 obtained by Belhafaf et al. [7].

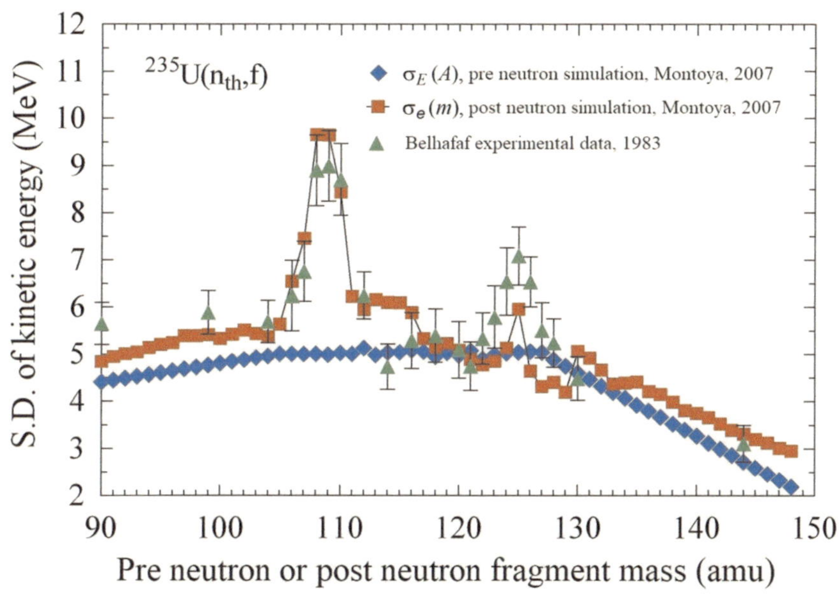

Figure 3.
Standard deviation of the final fragment kinetic energy distribution as a function of the final mass (m) of fragments from fission of ^{235}U induced by thermal neutrons. The output of Monte Carlo simulation (squares) [8] and experimental result (triangles) reproduce the experimental peaks obtained by Belhafaf et al. [7].

^{239}Pu(n$_{th}$, f), respectively [9]. For each of both reactions, those authors obtained a peak near the symmetric mass region in the corresponding σ_{tke} curves. For the reaction ^{235}U(n$_{th}$, f), the peak on σ_e around $\mu = 109$ reaches the value of approximately 6 MeV. This value must be compared to the 9 MeV found by Brissot et al. for σ_e around $= 109$. These authors used the LOHENGRIN mass spectrometer.

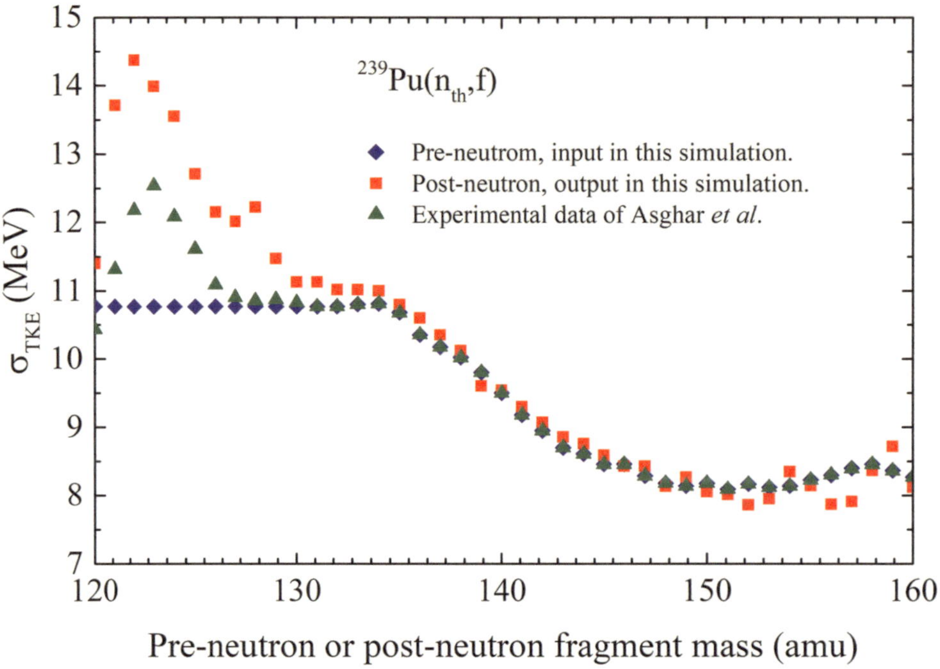

Figure 4.
Standard deviation of the total kinetic energy distribution as a function of the primary fragment mass (diamonds) and the corresponding to the provisional mass (squares), which are the input and output data, respectively, in a Monte Carlo simulation [10], to be compared with experimental data (triangles) obtained by Asghar et al. [9].

In 2020, simulating a primary distribution with any peak on the σ_{TKE} curve for fragments from the reaction ^{239}Pu(n_{th}, f), Montoya reproduced the values of σ_{tke} as a function of μ [10], found by Asghar et al. [9]. See **Figure 4**.

The Monte Carlo simulations of fission experiments show that the structures on the standard deviation of the fragment kinetic energy distribution, as a function of mass, are the result of prompt neutron emission and the combined effects of mass yield and average kinetic energy curves. The shape of those effects depends on the measurement technique.

A representative case is the region of light fragment near the symmetry of fission. In this region for both cases $\overline{E}(A), Y(A)$ are decreasing functions, and the final mass, due to neutron emission, is lower than the primary mass, i.e., $m < A$. As a consequence from that, for a given m, the final kinetic energy distribution is wider than the corresponding to the primary mass $A = m$.

In the same mass region, $n > n'$. Then, from the conservation relations, one can easily show that $\mu > A$. As a result of that, the distribution of e values for a given μ is different compared to the corresponding to the final mass m [11].

3. Average prompt neutron multiplicity

One of the quantities related to the distribution of fragment excitation energy in fission is the average prompt neutron multiplicity $\overline{\nu}(A)$ as a function of the mass. In

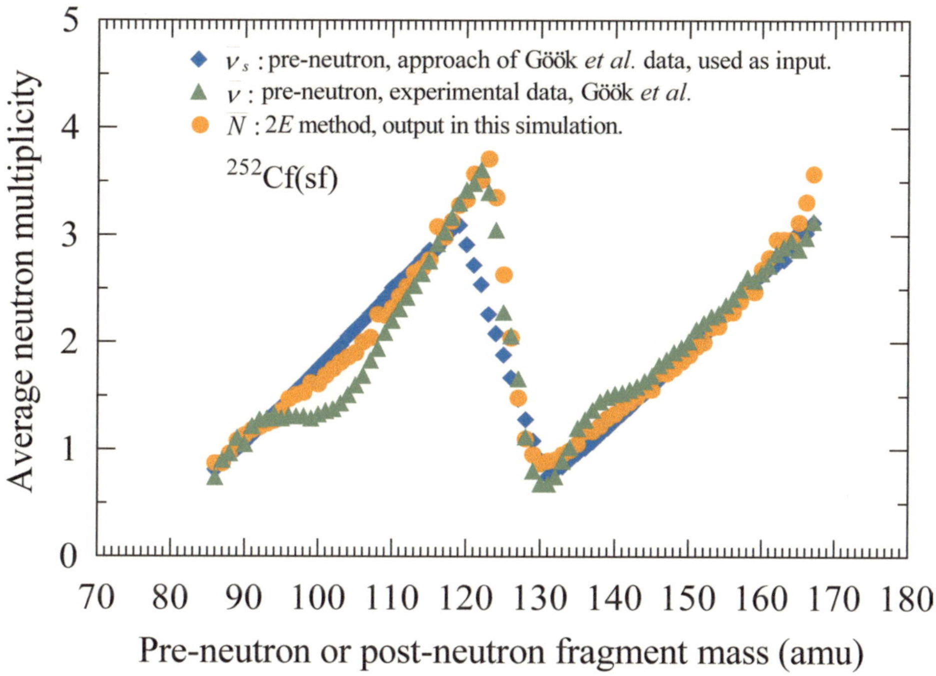

Figure 5.
Experimental average prompt neutron multiplicity as a function of mass (triangles) of fragments from the spontaneous fission of ^{252}Cf obtained by A. Göök et al. using the 2E method [12], its sawtooth approach of primary distribution (diamonds), and the corresponding to the calculated by a Monte Carlo simulation (circles) [11].

2014, using the 2E method, Göök *et al.* measured the average prompt neutron multiplicity as a function of the mass of fragments from the spontaneous fission of ^{252}Cf [12]. They found the well-known sawtooth approach, with a noticeable enhancement for $\mu = 121$. In 2019, applying a Monte Carlo simulation with input data representing a pre-neutron sawtooth approach ($\overline{\nu}_s$) without any enhancement for the average prompt neutron multiplicity, Montoya and Romero reproduced [11], as post-neutron, the Göök et al. result [12]. See **Figure 5**.

Those authors explain that the notorious peak in $\overline{\nu}(\mu)$ at $\mu = 121$ is an effect of the yield, the kinetic energy distribution, and the prompt neutron multiplicity as a function of primary fragment mass. In 2019, Montoya applied the Monte Carlo Method to reproduce the results of similar experiments for the ^{233}U(n_{th}, f) and ^{235}U(n_{th}, f) reactions [13].

Those studies show that to calculate the quantities distribution associated with the primary fragments it is necessary to apply the Monte Carlo technique to simulate the experiment and reproduce its results.

4. Yield of charge as a function of fragment kinetic energy

Several authors have measured the yield of charge as a function of the kinetic energy of the final light fragments from isobaric fragmentations. Quade et al. studied the reaction ^{233}U(n_{th}, f). The results show that the yield of the lower charge of light fragment mass increases as a function of the final kinetic energy [14].

Using the Monte Carlo method, Montoya and Rivera simulated a constant charge distribution as a function of the light fragment kinetic energy, the prompt neutron emission depending on the kinetic energy, and the yield of mass of primary fragments. Their results show that the increase in charge asymmetry with final kinetic energy is due to the decreasing neutron multiplicity as a function of the primary fragment kinetic energy [15]. See **Figure 6**.

Lang et al. obtained similar results for the reaction $^{235}U(n_{th}, f)$ [16] which were interpreted by Montoya and Rivera with similar arguments [17]. See **Figure 7**. The fragments with low kinetic energy have high excitation energy so they emit a high number of prompt neutrons. Thus, for a given final mass m, the corresponding primary mass and charge would be high. This reasoning is valid in a region where the yield is an increasing function of mass.

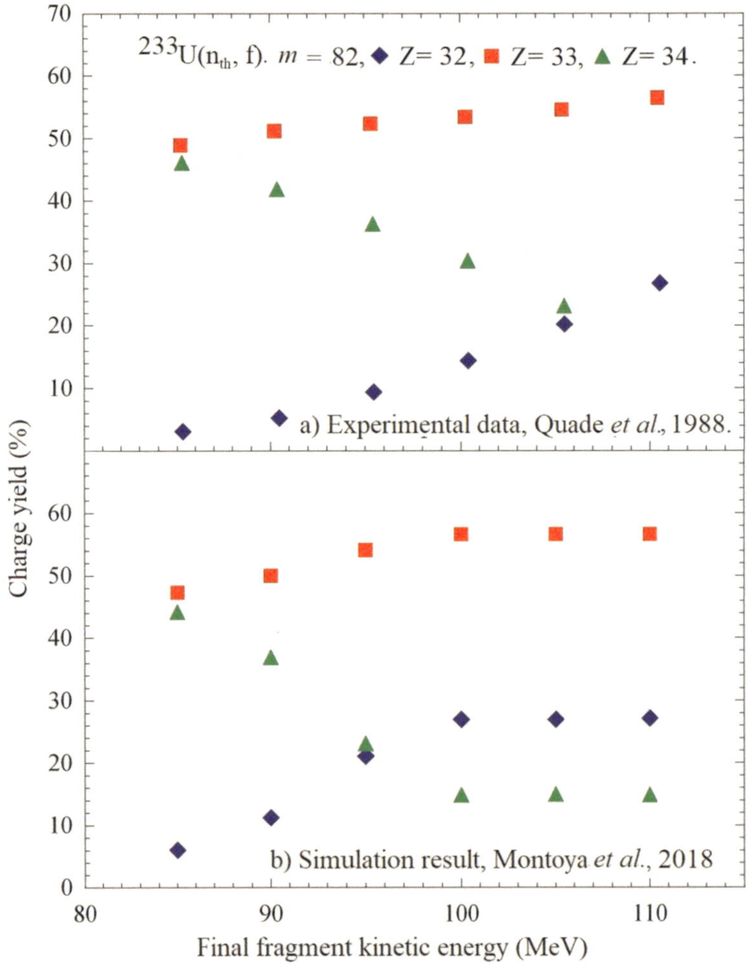

Figure 6.
Reaction ^{233}U (n_{th}, f). (a) The yield of charge as a function of final fragment kinetic energy for final mass 82 measured by Quade et al. [14]. (b) Results from Monte Carlo simulation of the experiment. Yields of charge lower than 0.1% are neglected [15].

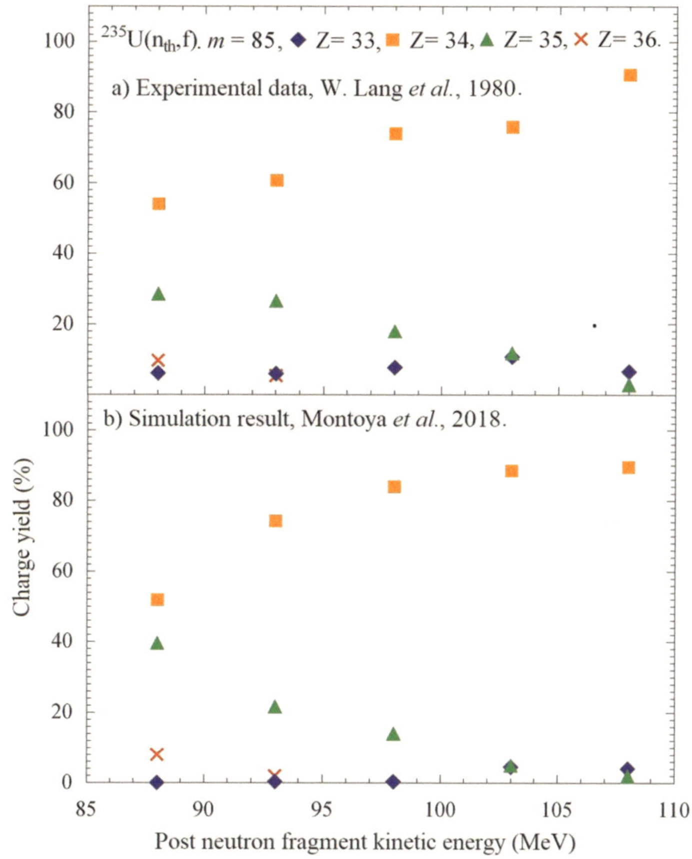

Figure 7.
Reaction $^{235}U(n_{th}, f)$. The yield of charge as a function of kinetic energy for mass number 85 of final fragments (after neutron emission). (a) Results measured by Lang et al. [16]; (b) results from Monte Carlo simulation of the experiment [17].

5. Cold fission

5.1 Maximum values of the kinetic energy of the fragments

Due to the emission of neutrons, the fragments reach the detectors with final values of mass and kinetic energy lower than the primary values. In 1981, Signarbieux *et al.*, discovered the phenomenon of cold fission in the reactions $^{233}U(n_{th}, f)$, $^{235}U(n_{th}, f)$, and $^{239}Pu(n_{th}, f)$ [18]. They detected fragments with high values of kinetic energy and, therefore, with low values of excitation energy, too low to emit neutrons. Using the time-of-flight difference technique, these authors separated neighboring masses, which permitted them to measure the maximal total kinetic energy curves as a function of the primary fragment mass ($TKE_{max}(A)$) [19]. See **Figures 8–10**, respectively.

In 1986, Trochon et al., using twin ionization chambers, obtained experimental values of $TKE_{max}(A)$ for the reaction $^{235}U(n_{th}, f)$ [20]. See **Figure 11**. The trends of

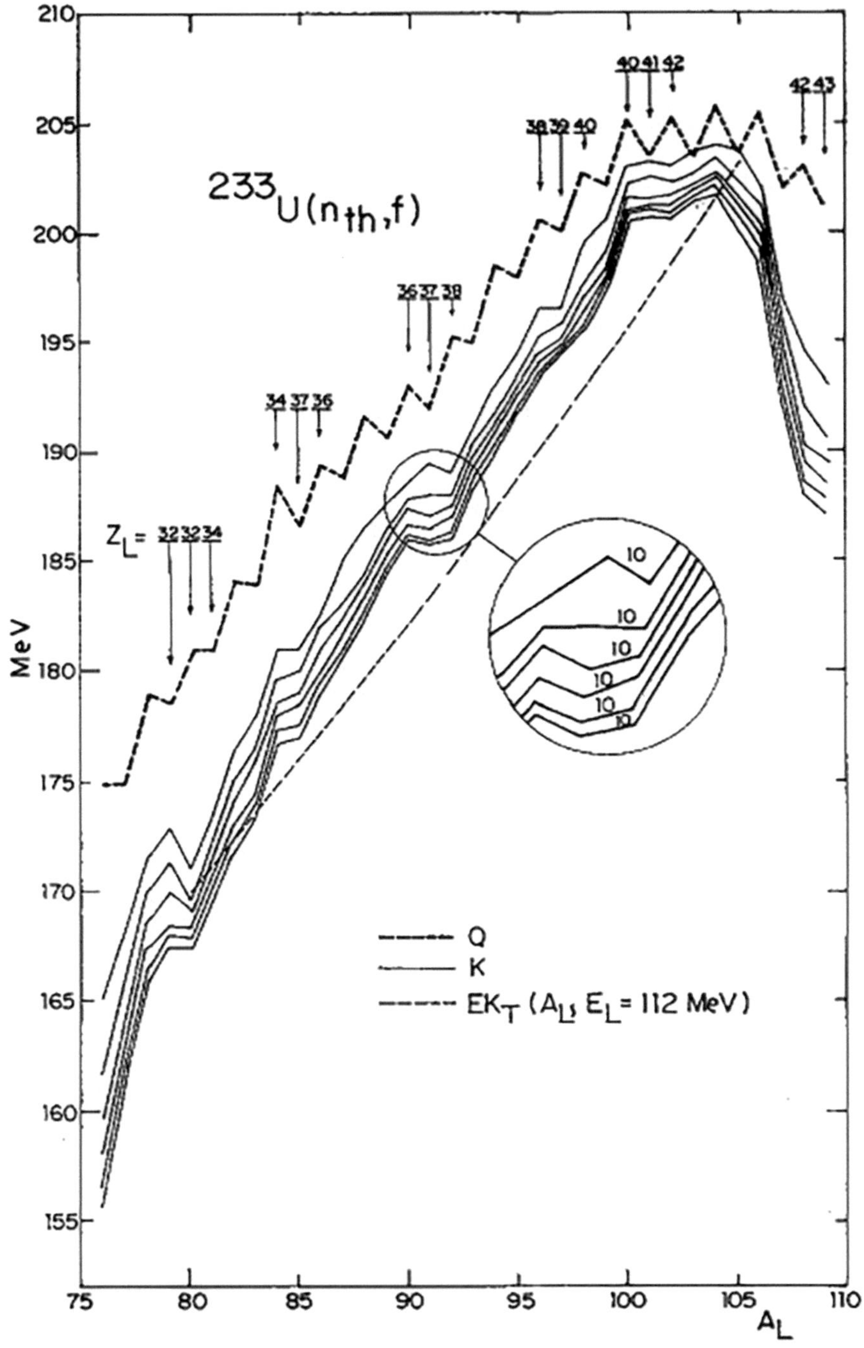

Figure 8.

Fragment mass and total kinetic energy distribution in cold fission region from $^{233}U(n_{th}, f)$. The thick dashed line is the maximal value of the available energy (Q), as a function of light fragment mass and its corresponding charge. The thin dashed line is the total kinetic energy as a function of fragment mas, for the light fragment kinetic energy $E = 112$ MeV. The thin full lines are equal probability lines: The numbers of detected fission events per a.m.u., whose corresponding total kinetic energy values between two consecutive lines, are indicated in the amplified circle. Those numbers (10 in this case) are chosen such that the separation between consecutive lines is of the order of 1 MeV. Figure is taken from [19].

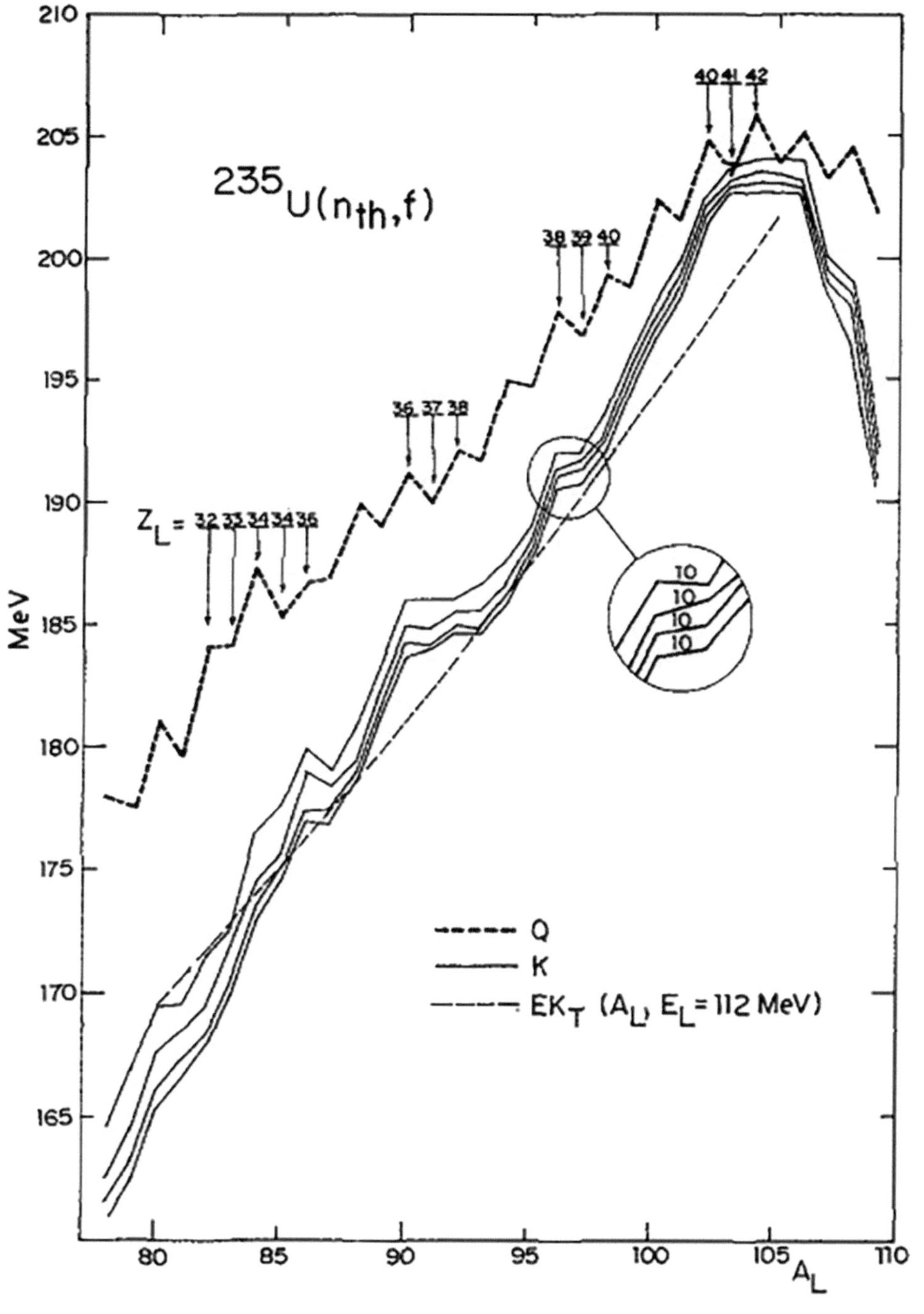

Figure 9.
Similar to ***Figure 6*** *for the reaction* $^{235}U(n_{th}, f)$. *Figure from [19].*

$TKE_{max}(A)$ are similar to the $\overline{tke}(\mu)$ curve, but approximately 20 MeV higher than those values [21–23]. The $TKE_{max}(A)$ curves show oscillations that were interpreted as effects of the change of fragment charges that reach these total kinetic energy values.

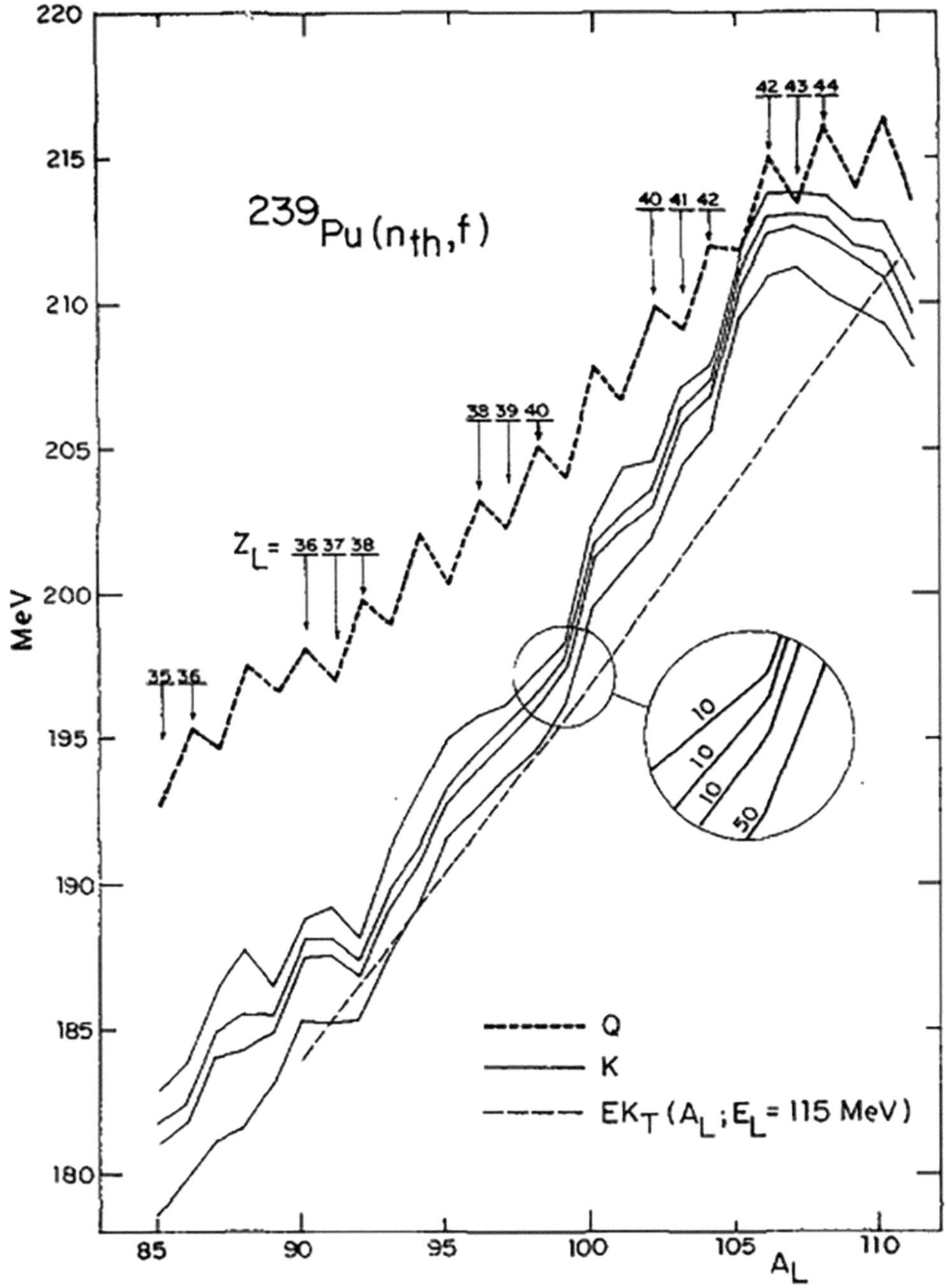

Figure 10.
*Similar to **Figure 7** for the reaction $^{239}Pu(n_{th}, f)$. Figure from [19].*

These so-called Coulomb effects produce ledges in the increasing part of $TKE_{max}(A)$ curve, corresponding to the change of charges that maximize the total kinetic energy [22].

The Coulomb effect is defined as the preference for a more asymmetric charge isobaric fragmentations corresponding to similar Q-values at the highest values of

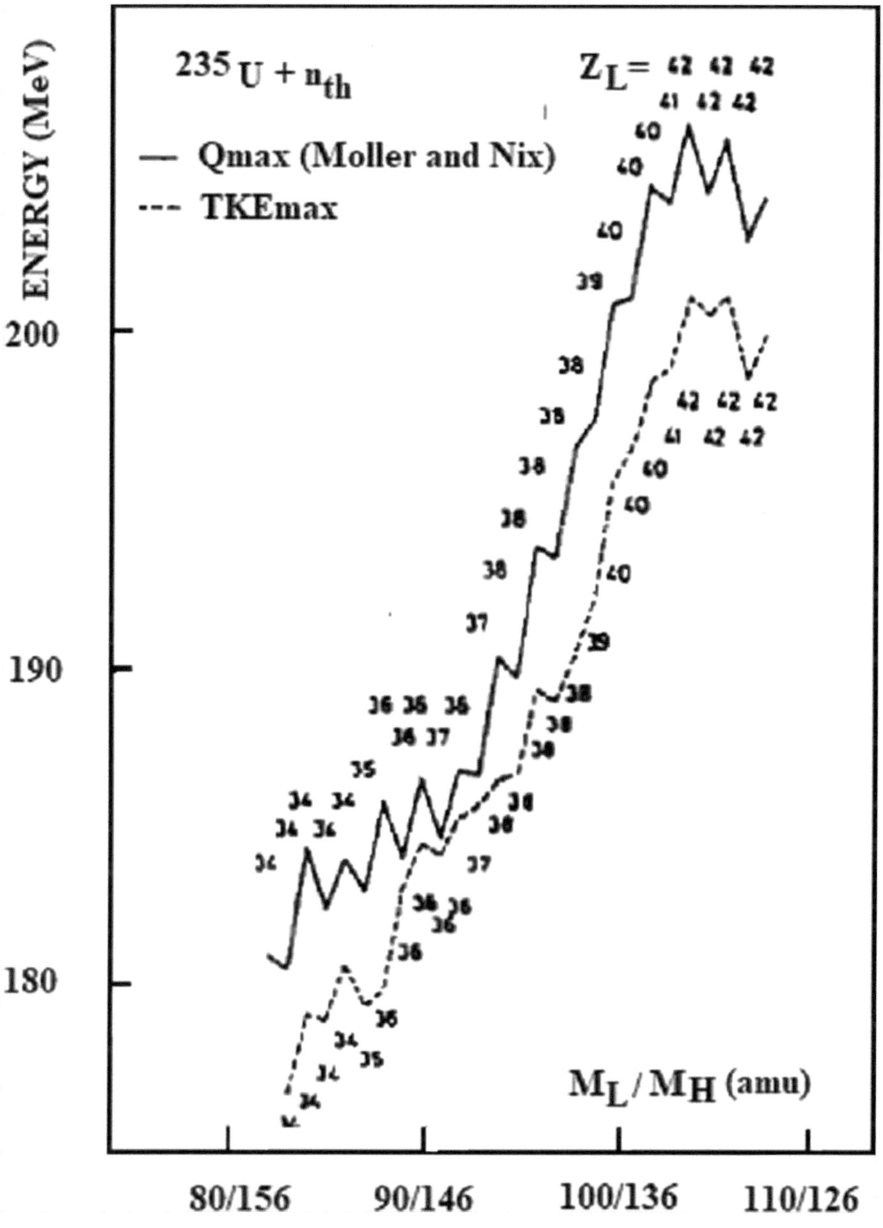

Figure 11.
Maximal value of total kinetic energy as a function of mass and its corresponding charge of fragments from thermal neutron-induced fission of ^{235}U. Figure taken from [20].

total kinetic energy [22]. Lang et al. [16] measured the charge and the mass distribution for $E = 108$ MeV. This energy window corresponds to the *TKE*-line 10 MeV lower than the average than the maximal Q-line. For the mass fragmentation 91/135, the charges $Z = 36$ and 37 correspond to the same Q-value (187.5 MeV), but the charge $Z = 36$ has a yield (75.9 ± 3.3%) higher than the corresponding to $Z = 37$ (19.7 ± 3.1%). For the same mass fragmentation, Trochon et al. [20] have deduced that the surviving charge at very high TKE-values is $Z = 36$. The same holds for the even-even fragmentations $(Z/Z', N/N') = (36/56, 56/88)$ and $(38/54, 54/90)$, whose Q values are 189.8 and 189.3, respectively.

5.2 Spherical-prolate configurations

The highest values of $TKE_{\max}(A)$ are reached for the configurations corresponding to the double magic spherical heavy fragments $Z = 50$ and $N = 82$ and the prolate and soft transitional light nuclei $Z = 40 - 42$ and $N = 60 - 64$. These conditions allow that

$$TKE_{\max}(A) \cong CE \cong Q, \tag{15}$$

Where CE is the Coulomb energy at the scission point, and Q is the energy available from the corresponding fragmentation.

Because in the symmetric mass region $CE \gg Q$, in the scission point, the fragments must be highly deformed to constitute a probable configuration. That implies a high deformation energy and therefore a low kinetic energy. This hypothesis is verified in the curve $TKE_{\max}(A)$: There is a steep drop in $TKE_{\max}(A)$ in symmetric fission [22].

5.3 Even-odd effects in cold fission

The curve of $TKE_{\max}(A)$ does not present clear even-odd effects as a function of the mass, contrary to what might have been expected in the case of a superfluid fission process [23]. This is not contradictory to the existence of even-odd effects in the number of protons δZ and δN. If there is no more than one pair of nucleons (protons or neutrons), the relationship between the three even-odd effects is [24, 25]:

$$1 + \delta A = \delta Z + \delta N. \tag{16}$$

In 1991 Gönnenwein and Börsig [26] found a negative odd–even effect on the minimum excitation energy ($TXE_{\min}$) as a function of Z. In 1993, Hambsch et al. [27] questions the positive odd even effect on that quantity. In 2013, Gönnenwein confirms the negative odd-even effect [28]. In 2016, Mirea proposes a microscopic model to explain the negative odd-even effects on excitation energy in cold fission [29]. In 2017, Montoya and Collin show that there is no contradiction between the positive odd-even effect on $TKE_{\max}$ and negative odd-even effect on $TXE_{\min}$ as a function of Z [30].

6. Conclusion

The emission of prompt neutrons from fragments from actinides fission does not permit to obtain the pre-neutron mass (A) and kinetic energy (E) distribution. It is only possible to measure the post-neutron mass (m) and kinetic energy (e). The difference between the post-neutron and pre-neutron distributions depends on the prompt neutron multiplicity (ν) associated with each complementary fragment. Using a Monte Carlo simulation of neutron emission and the experimental technique, several surprising unexpected results of the fission experiment are explained.

The curve of the standard deviation of the kinetic energy as a function of the final mass$\sigma_e(m)$ of fragments from the reaction $^{235}U(n_{th}, f)$ shows a peak around $m = 109$.

The curve of the average prompt neutron multiplicity as a function of the provisional mass of fragments, $\overline{\nu}(\mu)$, from the spontaneous fission of ^{252}Cf, presents a peak around $\mu = 122$. The results of Monte Carlo simulations show that both cases may be explained by the interplay of the prompt neutron emission and the rapid decreasing of $Y(A)$ and $\overline{E}(A)$ in their respective mass region.

For isobaric fragmentation in the asymmetric region, the preference for the lower charge of the light fragment increases with the final kinetic energy. This result is explained by the fact that fragments with higher kinetic energy have emitted a lower number of neutrons, which correspond to a lower charge nuclei.

In the cold fission region, the fragment excitation energy is so low that there is no emission of neutron emission. Thus, by a mass spectrometer, one can separate neighboring masses. So, we can obtain the curve of the maximal value of the primary total kinetic energy ($TKE_{\max}$) as a function of the fragment mass (A). The curve of the $TKE_{\max}(A)$ shows odd-even, shell, and Coulomb effects.

As a conclusion one may say that, in order to relate primary and final fragment distributions, it is necessary to simulate the primary fragment distribution, the emission of neutrons, and the experimental technique. Two different experimental techniques may produce contradictory results.

Author details

Modesto Montoya[1,2]

1 National University of Engineering, Lima, Perú

2 Ricardo Palma University, Lima, Perú

*Address all correspondence to: mmontoya@uni.edu.pe

References

[1] Hahn O, Strassmann F. Nachweis der Entstehung aktiver Bariumisotope aus Uran und Thorium durch Neutronenbestrahlung; Nachweis weiterer aktiver Bruchstücke bei der Uranspaltung. Naturwissenschaften. 1939; **27**(6):89-95. DOI: 10.1007/BF01488988

[2] Meitner L, Frisch OR. Products of the fission of the uranium nucleus. Nature. 1939;**143**(3620):471-472. DOI: 10.1038/143471a0

[3] Von Halban H, Joliot F, Kowarski L. Liberation of neutrons in the nuclear explosion of uranium. Nature. 1939; **143**(3620):470-471. DOI: 10.1038/143470a0

[4] Bohr N, Wheeler JA. The mechanism of nuclear fission. Physics Review. 1939; **56**(5):426-450. DOI: 10.1103/PhysRev.56.426

[5] Pomorski K, Dudek J. Nuclear liquid-drop model and surface-curvature effects. Physical Review C. 2003;**67**(4):044316. DOI: 10.1103/PhysRevC.67.044316

[6] Brissot R et al. Kinetic-energy distribution for symmetric fission of 236U. Proceedings 4th Symposium on Physics and Chemistry of Fission; Jülich, 1979. Vol. 2. Vienna: IAEA; 1980. pp. 99-110

[7] Belhafaf D et al. Kinetic energy distributions around symmetric thermal fission of U234 and U236. Zeitschrift für Physics: A Atomspheric Nucleus. 1983; **309**(3):253-259. DOI: 10.1007/BF01413757

[8] Montoya M, Saettone E, Rojas J. Monte Carlo simulation for fragment mass and kinetic energy distributions from the neutron-induced fission of 235U. Revista Mexicana de Fisica. 2007; **53**(5):366-370

[9] Asghar M, Caïtucoli F, Perrin P, Wagemans C. Fission fragment energy correlation measurements for the thermal neutron fission of 239Pu and 235U. Nuclear Physics, Section A. 1978; **311**(1–2):205-218. DOI: 10.1016/0375-9474(78)90510-9

[10] Montoya M. Oversize of the average prompt neutron multiplicity measured by the 1V1E method in the symmetric region of thermal neutron-induced fission of 239Pu. Results Physics. 2020;**17**:103053. DOI: 10.1016/j.rinp.2020.103053

[11] Montoya M, Romero C. Correlation between the average prompt neutron multiplicity as a function of the primary mass and the corresponding experimental mass of fragments from 252Cf spontaneous fission. Results Physics. 2019;**15**:102685. DOI: 10.1016/J.RINP.2019.102685

[12] Göök A, Hambsch F-J, Vidali M. Prompt neutron multiplicity in correlation with fragments from spontaneous fission of Cf 252. Physical Review C. 2014;**90**(6):064611. DOI: 10.1103/PhysRevC.90.064611

[13] Montoya M. Behavior of the average prompt neutron multiplicity as a function of post-neutron fragment mass in correlation with the pre-neutron fragment mass distribution in thermal neutron induced fission of 235U and 233U. Results Physics. 2019;**14**:102356. DOI: 10.1016/j.rinp.2019.102356

[14] Quade U et al. Nuclide yields of light fission products from thermal-neutron induced fission of 233U at different kinetic energies. Nuclear Physics A. 1988;**487**(1):1-36. DOI: 10.1016/0375-9474(88)90127-3

[15] Montoya M, Rivera A. Influence of neutron emission on the charge

distribution of final fragments from thermal-neutron-induced fission of 233U. Results Physics. 2018;**11**: 449-451. DOI: 10.1016/J.RINP.2018.09.039

[16] Lang W, Clerc HG, Wohlfarth H, Schrader H, Schmidt KH. Nuclear charge and mass yields for 235 U(n th , f) as a function of the kinetic energy of the fission products. Nuclear Physics A. 1980. DOI: 10.1016/0375-9474(80)90411-X

[17] Montoya M, Rivera A. Influence of neutron emission on the charge, mass and kinetic energy distribution of final fragments from 235 U(n th , f) reaction. Journal of Physical Communication. 2018;**2**(8):085016. DOI: 10.1088/2399-6528/aac07b

[18] Signarbieux C et al. Evidence for nucleon pair breaking even in the coldest scission configurations of ^{234}U and ^{236}U. Journal of Physics Lettres. 1981;**42**(19). DOI: 10.1051/jphyslet:019810042019043700

[19] Montoya M. Mass and kinetic energy distribution in cold fission of233U,235U and239Pu induced by thermal neutrons. Atomspheric Nucleus. 1984;**319**(2):219-225. DOI: 10.1007/BF01415636

[20] Trochon J, Simon G, Behrens JW, Brisard F, Signarbieux C. Cold fragmentation in thermal neutron induced fission of 235 U. Radiation Effects. 1986;**92**(1–4):327-331. DOI: 10.1080/00337578608208344

[21] Nishio K, Nakashima M, Nakagome. Multi-parametric measurement of prompt neutrons and fission fragments for 233U(n th, f). Journal of Nuclear Science and Technology. 1998;**35**(9). DOI: 10.1080/18811248.1998.9733919

[22] Montoya M, Hasse RW, Koczon P. Coulomb effects in low energy fission. Zeitschrift für Physics. 1986;**325**(3): 357-362. DOI: 10.1007/BF01294620

[23] Signarbieux C. Cold fragmentation properties: a crucial test for the dynamics of fission. International Workshop on Dynamical Aspects of Nuclear Fission. Smolenice (Czechoslovakia). 17-21 Jun 1991. Available online: https://inis.iaea.org/search/search.aspx?orig_q=RN:23073327

[24] Nifenecker H, Mariolopoulos G, Bocquet JP. Pair breaking mechanism in cold fission. Journal of Physics Letters. 1981;**42**(24):527-529. DOI: 10.1051/jphyslet:019810042024052700

[25] Montoya M. Fission: Viscosity and Odd-Even Effects

[26] Gönnenwein F, Börsig B. Tip model of cold fission. Nuclear Physics A. 1991; **530**(1):27-57. DOI: 10.1016/0375-9474(91)90754-T

[27] Hambsch F-J, Knitter H-H, Budtz-Jørgensen C. The positive odd-even effects observed in cold fragmentation—Are they real? Nuclear Physics A. 1993; **554**(2):209-222. DOI: 10.1016/0375-9474(93)90339-Y

[28] Gönnenwein F. Even-odd effects of fragment yields in low energy fission. Physics Procedia. 2013;**47**:107-114. DOI: 10.1016/J.PHPRO.2013.06.016

[29] Mirea M. Inversion of the Odd-Even Effect in Cold Fission from the Time-Dependent Pairing Equations. EPJ Web of Conferences. Vol. 122. 2016. p. 01009. DOI: 10.1051/epjconf/201612201009

[30] Montoya M, Collin V. "Positive Even-Odd Effects in the Maximal Kinetic Energy and Negative Even-Odd Effects in the Minimal Excitation Energy of Fragments from Thermal Neutron Induced Fission of 235 U". Revista Mexicana de Física. vol. 63. no.2. México. 2017

Section 3

Nuclear Technology

Chapter 4

Perspective Chapter: Assessment of Nuclear Sensors and Instrumentation Maturity in Advanced Nuclear Reactors

Thabit Abuqudaira, Pavel Tsvetkov and Piyush Sabharwall

Abstract

In the last decade, 97% of the worldwide commercial nuclear reactors connected to the grid were Light Water Reactors (LWRs). LWRs are expected to stay the dominant type of nuclear reactors for the next few decades. Reliable and redundant safety systems are required in nuclear reactors to ensure safe operation and shutdown in abnormal conditions. These safety systems are actuated by the signals obtained from several sensors and instrumentation in and out of the reactor core. In LWRs, these sensors and instrumentation have shown a high level of maturity with long operating experience. Ensuring the compatibility of these sensors and instrumentation with advanced nuclear reactors (Generation IV) is necessary. The compatibility of these contemporary technologies with advanced reactors was assessed by comparing the advanced reactors' environments with those of the currently operating reactors. In addition to that, the needed R&D for such technologies was highlighted. In comparison with the LWRs environment, it was shown that advanced reactor environments are expected to experience elevated temperatures, a fast neutron spectrum, and a harsh corrosion environment. It was demonstrated that R&D is required mainly for fixed in-core nuclear sensors and instrumentation, while it is not a priority for ex-core nuclear sensors and instrumentation.

Keywords: instrumentation and control, advanced reactors, reactor applications, extreme conditions, sensors

1. Introduction

According to the Power Reactors Information System (PRIS) developed by the International Atomic Energy Agency (IAEA) [1], Light Water Reactors (LWRs) share 84% of the nuclear reactors in operation worldwide. A total of 85% of these reactors are Pressurized Water Reactors (PWRs), whereas the remaining 15% are Boiling Water Reactors (BWRs). The remaining reactors in operation worldwide are Pressurized Heavy Water Reactors (PHWRs), Light Water Graphite-moderated Reactors (LWGRs), Gas-Cooled Reactors, Fast Breeder Reactors (FBRs), and

High-Temperature Gas-cooled Reactors (HTGRs). A pie chart showing the currently operating reactors classified by their type is shown in **Figure 1**.

Sixty-eight commercial nuclear reactors have been connected to the grid worldwide in the last decade. Sixty-four of these reactors were LWRs (PWRs). In the previous 5 years, construction was started in 30 nuclear reactors, 29 of them were also LWRs (PWRs). A pie chart showing the nuclear reactors under construction in the last 5 years classified by their type is shown in **Figure 2**.

In 2016, the first Generation III+ reactor was connected to the grid. Generation III+ reactors have been dominating the newly constructed reactors. It is anticipated that they will be leading for the next few decades. Generation IV reactors are expected to be deployed commercially within at least two decades [2]. With this trend of domination of LWRs, specifically PWRs, improvement, and development have to be continued for such types of reactors. On the other hand, the technological advancement in the next generation of reactors requires parallel progress in the R&D efforts for all needed components for their operation.

A nuclear reactor requires various redundant sensors and instrumentation systems to operate safely and reliably. These sensors and instrumentation vary in function and the degree of precision necessary to fulfill their purpose. The information gained from these sensors, and instrumentation is used as input for the reactor's control and

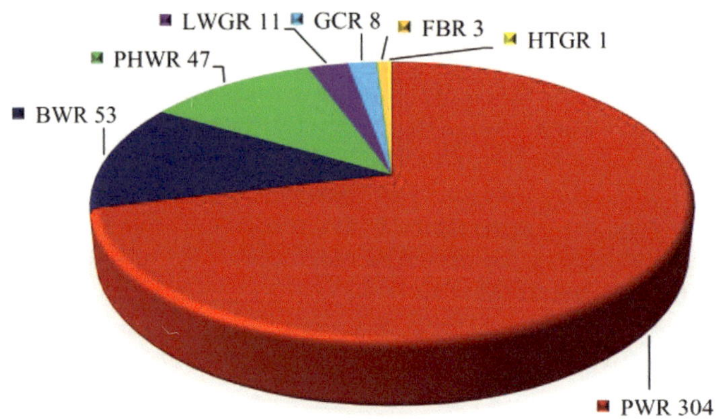

Figure 1.
The number of currently operating reactors by type (As of the end of 2022).

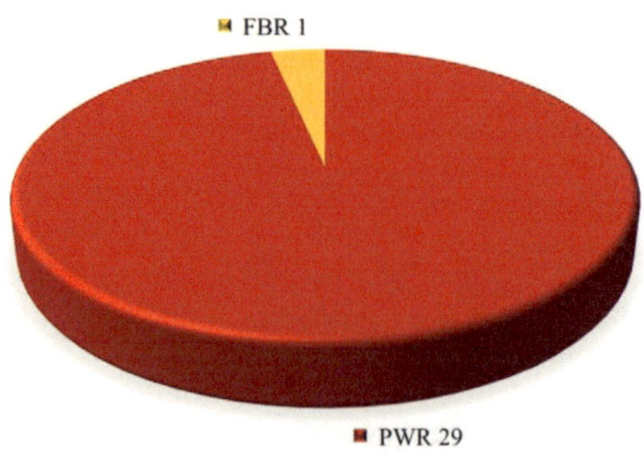

Figure 2.
The number of reactors under construction by type (As of the end of 2022).

to indicate transients and deviations [3]. Thus, in addition to their role as control systems, sensors and instrumentation play a role as safety systems.

Since the first experiment with a nuclear reactor (Chicago Pile-1), sensors and instrumentation were required to monitor and ensure the safe operation of the reactor. Afterward, the number of sensors and instrumentation needed for all nuclear reactors constructed tended to increase. This increment was mainly to improve the safety of the newly operated reactors. Research and Development (R&D) in sensors and instrumentation have gained extra attention following the accident at the Three Mile Island Unit-2 (TMI-2). Investigations were performed by several National Laboratories, private companies, and consultants for the Department of Energy (DOE) to assess the instrumentation and electric components' ability to accomplish their design function [4]. The accident provided an opportunity to evaluate sensors and instrumentation performance in severe conditions [5].

In LWRs, sensors and instrumentation passed a long way of R&D until they reached a high level of maturity. However, the advanced reactor designs pose new challenges to these sensors and instrumentation, mainly due to the differences in the operating environments of the reactors. Some sensors and instrumentation, particularly in-core sensors and instrumentation, will operate in conditions significantly different from those in the current generation of nuclear reactors. High-temperature and corrosive environments can impose significant challenges to the design of sensors and instrumentation. In addition to that, these operating conditions could also impact sensors and instrumentations' reliability and accuracy. As a result of these factors, R&D programs are emerging, intending to commercially obtain sensors and instrumentation specifically designed for these types of reactors.

Advanced reactor designs are experiencing rapid technological advances. For this reason, R&D in sensors and instrumentation technologies must be initiated. The United States Nuclear Regulatory Commission (U.S.NRC) addressed Congress that research is needed to develop and assess the performance of new types of sensors and instrumentation for advanced reactors [6]. The Department of Energy, Office of Nuclear Energy (DOE-NE) has established several projects to accelerate R&D in sensors and instrumentation technologies in the United States. One of these programs is the Light Water Reactor Sustainability (LWRS) program. The program focuses on improving the economics and reliability and sustaining the safety of the U.S.'s fleet of Nuclear Power Plants (NPPs) [7]. Modernization of instrumentation technology is one of the technical areas of R&D of the program.

Another program is the Nuclear Energy Enabling Technologies (NEET) program. Advanced Sensors and Instrumentation (ASI) is one of its elements [8]. ASI focuses on research to develop new types of sensors and instrumentation systems that will likely be used in existing and advanced reactors. The program began in 2011. R&D activities have been effectively supported through directed research and competitive awards, enabling significant advancements in the field.

Nuclear sensors and instrumentation are one of the main types of reactor sensors and instrumentation required to operate the reactor safely and efficiently. Various detectors measure neutron flux and reactor power in the currently operating reactors. It is expected that a new generation of such types of detectors is required to be able to work in advanced reactor environments. Thus, R&D has to be continued to modify and improve the currently available set of sensors and instrumentation and to develop the next generation of such technologies.

The first step toward developing new types of nuclear sensors and instrumentation is to assess the maturity of the currently available sensors and instrumentation.

Identifying the compatibility of the currently used sensors and instrumentation with the advanced reactor environment and analyzing the technological gaps and the needed improvements in the available technology will help to guide the R&D toward the most crucial areas. Such an assessment can help to meet the requirements of a new NPP license. For these reasons, an approach was developed to assess the R&D needs in nuclear sensors and instrumentation technologies tailored to future designs of nuclear reactors.

2. Nuclear reactor environments

Many reactor designs have been proposed since the first critical reactor, Chicago Pile 1. However, not all of them could achieve the stage of commercial operation. Few designs have reached the experimental stage, while many stayed as concepts. Differences in nuclear reactor designs led to significant differences in the operating environments. Thus, adopting a new reactor design into commercial operation faces many challenges, mainly due to the lack of operating experience with the new reactor environment. Understanding the differences in the operating conditions between the commercially available nuclear reactors and the proposed advanced reactor types will help to analyze the problems that sensors and instrumentation would face in advanced reactor environments.

2.1 Current and advanced reactor designs

In this study, the proposed Generation IV reactors provided by the Generation IV International Forum were adapted for comparison with the currently available LWRs [9]. These are the Supercritical Water Reactor (SCWR), the Very High-Temperature gas-cooled Reactor (VHTR), the Sodium-cooled Fast Reactor (SFR), the Lead-cooled Fast Reactor (LFR), the Gas-cooled Fast Reactor (GFR), and the Molten Salt reactor (MSR). This study did not include a comparison of the operating conditions of the currently operational PHWR and LWGR designs.

2.1.1 Pressurized water reactor

LWRs are thermal neutron spectrum reactors. They have gained their name since light "ordinary" water is used as a coolant and moderator [10]. Huge experience has been achieved in operating LWRs, including Pressurized Water Reactors (PWRs) and Boiling Water Reactors (BWRs). In the United States, all currently operating 93 reactors are LWRs. Sixty-two of them are PWRs, and 31 are BWRs [11]. This operating experience has given enough maturity to nearly all components required for a reactor operation.

Since the first PWR "Shippingport" constructed by Westinghouse [12], PWRs have dominated the worldwide nuclear reactors market. For this reason, PWRs have been considered a reference "Benchmark" in developing new reactor designs. In general, newly proposed advanced reactor designs are compared with PWRs to assess their advantages and benefits. In addition, most of the proposed LWR designs for the future technology of Small Modular Reactors (SMRs) are PWRs [13]. Several vendors of PWRs are available worldwide. AP1000 reactor design is selected to be a reference for the operating environment for PWRs [14]. All other PWRs designs have quite similar operational parameters.

2.1.2 Boiling water reactor

Unlike PWRs, light water in BWRs is permitted to boil within the reactor core. The steam from this boiling process is sent directly to the turbine without passing through a heat exchanger. Thus, BWRs operate at a lower pressure than PWRs (around 7 MPa). The maximum coolant temperature in the reactor core is around 288°C. In addition to that, the reactor has lower power density in comparison with PWRs.

Future designs for BWRs are significantly fewer than PWRs. In the United States, two BWR designs have been approved by the U.S.NRC [15]. These are the Advanced Boiling Water Reactor (ABWR) and the Economic Simplified Boiling Water Reactor (ESBWR). Furthermore, BWRs are considered for small modular reactor designs (SMR), such as the BWRX-300 design by GE Hitachi in the United States [16]. The development of these advanced BWRs resulted from many years of design improvements, drawing from the valuable experience gained through early reactor designs. For operating parameters comparison, the ABWR design was selected.

2.1.3 Supercritical water reactor

Supercritical Water-Cooled Reactor (SCWR) is a high-temperature, high-pressure, light water-cooled reactor that operates above the thermodynamic critical point of water (374°C, 22.1 MPa) [17]. Even though it uses light water as a coolant, SCWR is considered an evolution of both PWRs and BWRs. It is one of the proposed advanced reactor designs by the Generation IV International Forum (GIF). Unlike LWRs, SCWRs have designs to operate with a fast-neutron spectrum [18]. The United States is interested in SCWRs with a thermal neutron spectrum.

Currently, there is no SCWR under construction. The reactor is still in the phase of pre-conceptual design. Several reactor designs have been proposed worldwide. However, this type of reactor is expected to have meager chances for commercial development compared to other advanced reactor types. However, the share of research papers published on this type of reactor during 2004–2018 among different reactor types was around 9% [19]. The reactor parameters of the proposed U.S. thermal-spectrum SCWR are used as a reference for this study for comparison purposes [20].

2.1.4 Very high-temperature gas-cooled reactor

Various experimental and commercial gas-cooled reactor designs have been developed around the world. These designs are Dragon (United Kingdom), Peach Bottom (United States), Arbeitsgemeinschaft Versuchsreaktor - AVR (Germany), Fort St. Vrain (United States), Thorium High-Temperature Reactor - THTR (Germany), High-Temperature Engineering Test Reactor—HTTR (Japan), High-Temperature Reactor—HTR-10 (China), and the Advanced Gas-cooled Reactor—AGR (United Kingdom). This operating experience has given enough maturity to nearly all components required for reactor operation. However, advanced gas-cooled reactors would exhibit more challenging operating environments.

VHTR is an evolutionary proposed design of the previously mentioned reactors. It would have operating temperatures of around 1000°C. It uses helium as a coolant, with the fuel being tiny, coated fuel particles embedded in a graphite matrix [21]. Compared with LWRs, the reactor power density is significantly lower, while the coolant temperature is much higher. This type of reactor got attention since the produced heat can be used for other industrial purposes. In addition to that, the operating pressure considered for the VHTRs is comparable to BWRs.

2.1.5 Sodium-cooled fast reactor

In SFR, the molten sodium metal is used as the reactor coolant enabling the reactor to have a fast neutron spectrum. Many safety and economic features were the reason for selecting this reactor design as one of the proposed generation IV nuclear reactors. Several SFR have been built around the world. Currently, two reactors in operation in Russia are SFR. TerraPower plans to construct the Traveling Wave Reactor (TWR), a sodium-cooled reactor in the United States.

The technology for SFRs has mainly been established through previous fast reactor programs. SFR is one of the most promising reactor technologies to close the nuclear fuel cycle. The design of the SFR will be used as a breeding/burning uranium-plutonium fuel cycle [22]. The proposed Generation IV SFR was used as a reference to be compared with other reactor operating conditions [23]. It can be observed that the main feature of such a reactor is the reactor's high-power density in comparison with the conventional LWRs. In addition, sodium's high boiling point allows the reactor to operate at atmospheric pressure.

2.1.6 Lead-cooled fast reactor

A fast-neutron spectrum with lead or lead/bismuth as the coolant characterizes the LFR. Lead as a coolant has several advantages compared to other reactor coolants. The coolant's high boiling point provides a large margin for safety in preventing coolant boiling and maintaining core integrity during accidents. Thus, many simplifications in the reactor design improved the overall economic performance. Compared with LWRs, the reactor operates at atmospheric pressure, and the molten lead coolant is at higher temperatures.

Several countries, including China, Russia, the United States, Sweden, Korea, and Japan, actively develop various LFR concepts. The LFR system may be the first Reactor of Generation IV to be deployed. The first lead-cooled reactor construction started in 2021.

2.1.7 Gas-cooled fast reactor

The GFR features a fast-neutron spectrum with helium as the coolant [24]. Thus, in addition to the increased operating temperature in such reactors as the VHTR, it has a fast neutron spectrum like SFR. The GFR design aims to operate safely and reliably, achieving sustainability, economic competitiveness, and enhanced proliferation resistance and security [25]. In addition, being a fast reactor will allow the usage of recycled plutonium, which will play a crucial role in the long-term strategy for nuclear power development [26].

The proposed Generation IV GFR design is a 2400 MW_{th} plant operating with a coolant outlet temperature of 850°C. The power density of the core is comparable to that in PWRs, while the coolant pressure is comparable to the BWRs. GFR technology is still in the R&D stage. Moreover, the design represents a significant step toward developing large-scale, high-temperature gas-cooled fast-spectrum reactors.

2.1.8 Molten salt reactor

MSRs come in two distinct designs regarding fuel geometry. The first design is when the fuel and the coolant are mixed in one homogeneous fluid. In the second one, the fuel is confined to traditional fuel rods. These two designs have some differences in the operating environment. This study uses the salt-fueled design as a reference for the comparison. The fuel movement, salt chemistry, higher power density, and elevated temperatures significantly differ from the currently operating reactors.

In the United States, liquid fuel nuclear reactors got an attraction in the 1950s by initiating the Molten Salt Reactor Program (MSPR) at Oak Ridge National Laboratory (ORNL). Two MSRs were built and operated: the Aircraft Reactor Experiment (ARE) and the Molten Salt Reactor Experiment (MSRE) [27, 28]. The ARE was successfully operated for over 4 days, while the MSRE was for 4.5 years [29]. Nevertheless, no commercial deployment of this reactor type was achieved.

In the United States, research is conducted on molten salt reactor development and deployment efforts. Construction is underway for a Molten Salt Research Reactor (MSRR) located on the Abilene Christian University (ACU) campus in Texas. The primary purpose of the Reactor will be to facilitate research and provide training opportunities within the Molten Salt Reactor (MSR) environment.

2.2 Comparison of reactor environments

Coolant temperature and pressure, neutron spectrum, and corrosive environment are the most important parameters regarding the compatibility of new sensors and instrumentation technologies with an advanced reactor environment. Although sensors and instrumentation that would be placed in-core may exhibit slightly higher temperatures than coolant outlet temperatures, this temperature can indicate the temperature range in which the sensors and instrumentation should be able to operate. The operating pressure measurement is necessary to prevent instrumentation failures, damage, or compromised performance. The knowledge of the type of neutron spectrum in a particular reactor indicates the radiation damage and material degradation with time. The coolant material describes the corrosion environment that may face the sensors and instrumentation. A summary of the operating parameters that may make challenges to the currently available sensors and instrumentation in current and advanced reactors is listed in **Table 1**.

Table 1 shows that advanced nuclear reactors will operate at temperatures higher than what sensors and instrumentation have experienced serving in currently available nuclear reactors. In addition, higher levels of power density, fast neutron spectrum, and the corrosive environment in these reactors may pose a challenge for the available sensors and instrumentation.

In terms of neutron flux, the overall neutron flux level will be comparable for all reactor designs. However, only the type of neutron spectrum and the activation of the coolant material and the associated gamma-ray field may play a role in the compatibility of the sensors and instrumentation. Studying each parameter's impact on the compatibility of the sensors and instrumentation can indicate the effect of this parameter on all advanced reactor designs. The proposed advanced reactor types are compared with LWRs in terms of the operating parameters that may affect the sensors and instrumentations. This comparison is shown in **Table 2**.

Reactor type (Design)	Neutron spectrum	Coolant type	Coolant outlet temperature (°C)	Pressure (MPa)	Power density (MW_{th}/m^3)
PWR (AP1000)	Thermal	Light water	330	15.5	98
BWR (ABWR)	Thermal	Light water	288	7.07	49.2
SCWR*	Thermal	Light water	500	25	97.6
VHTR*	Thermal	Helium	1000	7	6
SFR*	Fast	Sodium	550	0.1 (Ambient)	350
LFR*	Fast	Lead	480	0.1 (Ambient)	69
GFR*	Fast	Helium	850	7	100
MSR*	Fast	Sodium	700	0.1 (Ambient)	500

**Designs provided by the Generation IV International Forum.*

Table 1.
Comparison of operating parameters for different types of reactors.

Reactor type	Operating parameters (In comparison with LWRs)			
	High temperature	Corrosive environment	High pressure	Fast spectrum
SCWR	✓	x	✓	x*
VHTR	✓	x	x	x
SFR	✓	✓	x	✓
LFR	✓	✓	x	✓
GFR	✓	x	x	✓
MSR	✓	✓	x	✓

**Some SCWRs are available with fast spectrum design.*

Table 2.
Advanced reactors environments in comparison with LWRs.

It can be concluded that advanced reactor environments will differ from the currently available LWRs by the increase in the operating temperature (from the range of 300–350°C up to around 1000°C in VHTR), the harsh chemical environment caused by the usage of molten salts and liquid sodium and lead, and the exposure to fast neutrons in fast reactors. In addition, it can be demonstrated that high pressure is not a significant concern in advanced reactor designs. Only the SCWR design has an elevated operating pressure than the currently available reactor designs.

3. Assessment methodology

The previous section discussed the main differences in the operating environment of several nuclear reactors. Commercially available nuclear reactors have attained enough operating experience with various sensors and instrumentations. Thus, the first step to discovering the problems that sensors and instrumentation would face in advanced reactors is to study the differences in the operational environments between the commercially available reactors and the proposed advanced reactor.

An approach was adopted to investigate nuclear sensor and instrumentation technology's R&D needs. This approach analyzed the available nuclear sensors and instrumentation used in current reactor designs. The technological gaps and the needed improvements were highlighted. Then, the compatibility of these sensors and instrumentation with the advanced reactor environments was investigated.

Sensors and instrumentation that would not face any problem in the advanced reactor environment and are commercially available and can be used in advanced reactor designs. Thus, the instrumentation technology can be assessed as mature. Therefore, R&D is not a priority. If the sensor and instrumentation technology is compatible with the advanced reactor environment but is not commercially available, an R&D program is essential to commercialize the technology. Commercial availability of several types of sensors and instrumentation can be investigated by searching for the availability of the technology within companies of the Nuclear Suppliers Association (NSA). NSA comprises approximately 73 companies specializing in manufacturing and distributing services and products for nuclear energy users [30].

Sensors and instrumentation facing problems in the advanced reactor environment require significant R&D programs to fill the technological gaps. In either case, where R&D is needed based on the defined approach, the availability of initiated research projects on the sensor and instrumentation technology was further investigated. The investigation was based on the availability of research projects supported by the DOE NEET ASI program. Although the availability of funded projects on a particular type of technology does not mean that the technology is mature, it indicates that sensors and instrumentation with no current research projects have priority for future R&D programs. The adopted approach for determining the need for R&D programs for specific instrumentation is summarized in **Figure 3**.

In the future development of this approach, the Technology Readiness Levels (TRLs) approach can be used to assess the maturity of a particular instrumentation technology. NASA first proposed the TRL in the 1980s. The TRLs are divided into nine technology readiness levels starting from TRL 1 (The lowest where the basic principles for a specific technology are observed and reported) and ending in TRL 9 (The highest where the existing system was proven to be successful for mission operations) [31].

4. Nuclear sensors and instrumentation

Various sensors and instrumentation are required in a nuclear reactor to operate it safely and efficiently. These sensors and instrumentations represent the information suppliers to the reactor operators about the reactor status. Based on their deployment location, they are needed for the reactor core or other nuclear reactor components. These include nuclear instrumentation required to monitor the neutron flux and the associated reactor power, process instrumentation to measure non-nuclear processes such as temperature distribution in the core, pressure and flow measurements of the coolant, radiation instrumentation systems for radiation monitoring and dose levels measurements, and some other specific instrumentation for specific purposes [32]. A summary of nuclear reactor sensors and instrumentation groups is shown in **Figure 4**.

Process instrumentation is crucial in monitoring and controlling various parameters within a nuclear reactor. It involves using sensors, detectors, and other measurement devices to gather data and provide real-time information about the reactor's

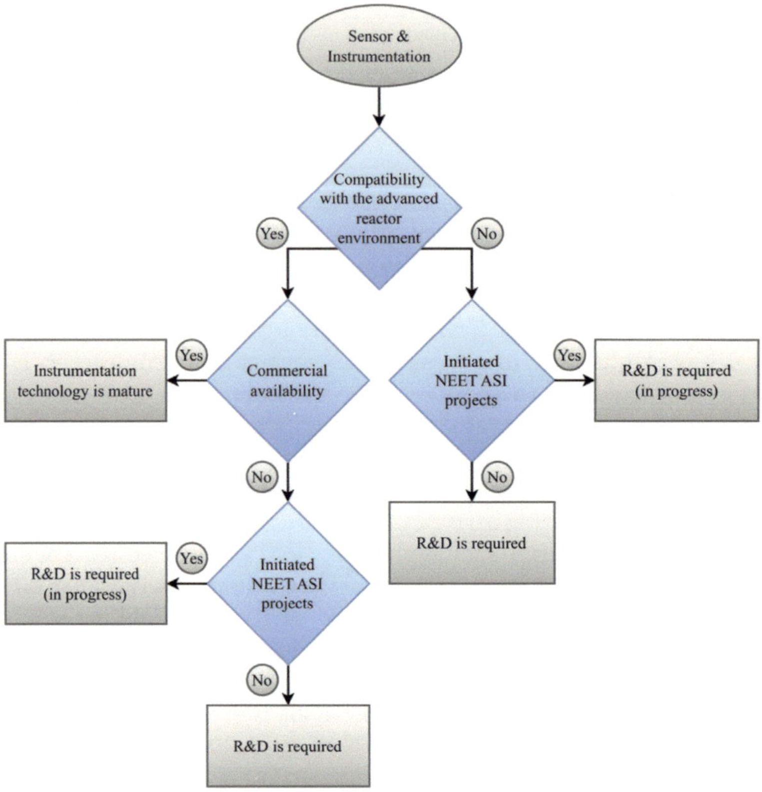

Figure 3.
Sensors and instrumentation maturity assessment approach.

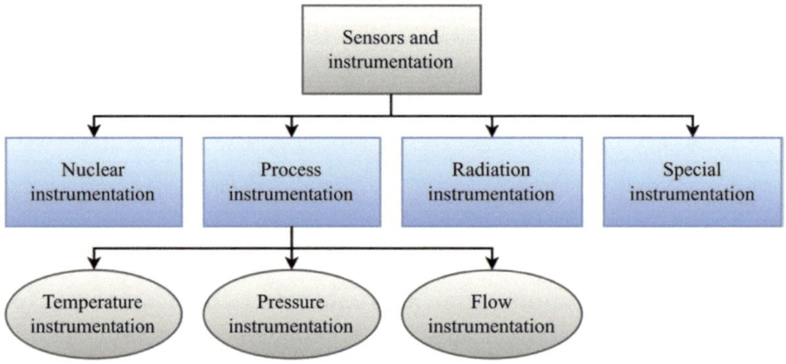

Figure 4.
Sensors and instrumentation groups of a nuclear reactor.

temperature, pressure, flow, and other essential parameters. Radiation instrumentation systems include the sensors and instrumentation required to monitor the site radiation level (area radiation monitoring), gas effluents, and main steam pipes. These instrumentation systems are essential to protect personnel from radiation exposure, and they can indicate abnormal reactor conditions. Special instrumentation systems are reactor-type targeted. Based on the reactor type, such as the boric acid concentration instrumentation in PWRs, these sensors and instrumentation are required.

Nuclear sensors and instrumentation play a significant role in reactor control and safety. They measure the neutron flux and the associated reactor power level during various reactor conditions. These are: during the approach to criticality and reactor startup, intermediate power operation, and full-power reactor operation [33]. In the Advanced Pressurized water-cooled reactor (AP1000) designed by Westinghouse Electric Company, several types of neutron detectors are used to monitor the neutron flux from zero to 120% of full power [34].

In the currently operating reactors, various neutron detection systems are being used. These sensors and instrumentation can be installed inside the reactor core (in-core) or outside the reactor core (ex-core) based on their design purpose. They provide diversity and redundancy in the obtained information about the reactor status. Fission chambers, Ionization chambers, gamma thermometers, proportional counters, and self-powered neutron detectors (SPNDs) are among this nuclear instrumentation [35]. Furthermore, other nuclear instrumentation systems are used in specific reactor types, such as the aeroball measurement system used in EPR reactors. In addition, research is ongoing to apply new reactor flux and power measurement methods, such as the Cerenkov monitoring systems for in-core power measurement. The typical flux ranges covered by nuclear instrumentations in LWRs are illustrated in **Figure 5** [33]. Note that the nuclear instrumentations should be able to overlap between the operating flux ranges.

A group of fixed in-core and ex-core neutron detectors in nuclear reactors monitors the neutron flux and reactor power. In AP1000 reactors, fixed in-core and ex-core neutron detectors are used to monitor the neutron flux and reactor power [36]. The choice of detector location is made based on the detector's purpose and ability to withstand the harsh environment in the reactor core. These in-core and ex-core nuclear instruments should be able to measure the neutron flux over the entire range from the source start-up range up to the full power range.

In-core instrumentation should be able to operate at the operating conditions of the nuclear reactor. Ex-core instrumentation would be exposed to lower temperature, pressure, and neutron and gamma flux values. In commercial nuclear reactors, self-powered neutron detectors and gamma thermometers are the in-core nuclear instrumentation. Ionization chambers and proportional counters are the ex-core nuclear instrumentation. Fission chambers are used as both in-core or ex-core nuclear instrumentation. Classification of nuclear instrumentation based on their location regarding the reactor core is shown in **Figure 6**.

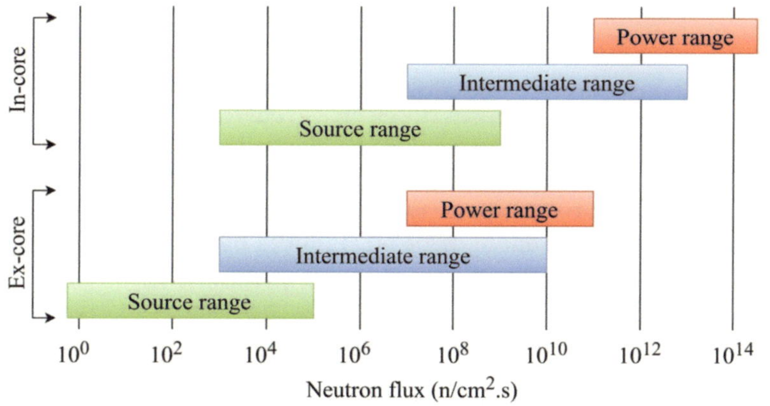

Figure 5.
The covered typical flux ranges by nuclear instrumentation systems in LWRs.

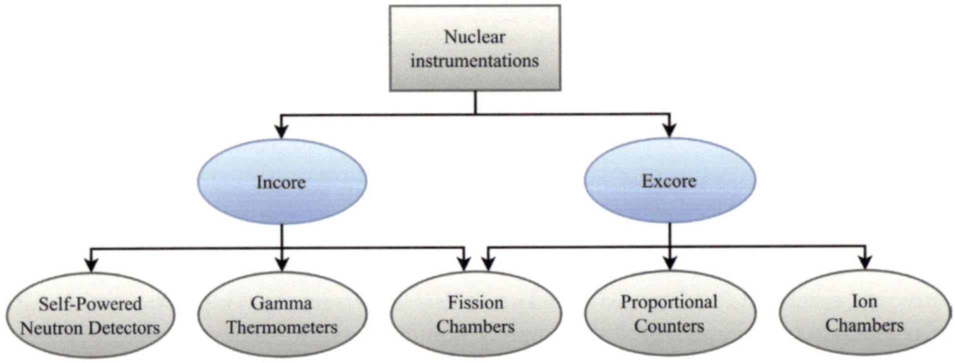

Figure 6.
Classification of nuclear instrumentation by deployment location.

Regarding the movability of the installed nuclear sensors and instrumentation, they can be fixed in the reactor core or movable on-demand systems. The aeroball measurement system is an example of a system used as on-demand in-core nuclear instrumentation.

4.1 Self-powered neutron detectors

Self-Powered Neutron Detectors (SPNDs) are the most common detectors used in commercial nuclear power plants [37]. These detectors attained this wide usage due to their advantages over other detector types in the reactor operating environments. Simple structure, tiny volume, no need for a power supply, and high temperature and radiation resistance are among the advantages of this type of detector [38].

In a nuclear reactor, several SPNDs are located axially within a channel in several channels inside the core. Thus, SPNDs can continuously measure the neutron flux in fixed positions inside the reactor core from low power level up to full power. However, they are unsuitable for startup range measurements. A SPND comprises four main components: emitter, insulator, collector (sheath), and lead wire [39]. A schematic drawing of a SPND is shown in **Figure 7**.

The emitter material is selected to have a high neutron absorption cross-section that undergoes betas decay directly or indirectly following the neutron absorption. The insulator's function is to provide temperature and electric resistance. It must have an optimized thickness so that the produced electrons in the emitter can reach the sheath (collector). The sheath works as a collector of the electrons from the emitter. The lead wire's function is to transmit the electrical current. A current meter will read the current value, which will be proportional to the emitted electrons from the emitter material. After equilibrium, the registered electric current will be proportional to the incident neutron flux from the reactor core. Thus, the reactor power is measured by measuring the output current in the SPND induced by the decay resulting from the neutron capture reactions [40].

The selection of the emitter material determines the response of the SPND. The response of an SPND is dependent on the radioactive decay of the radionuclide resulting from the neutron capture by the emitter material. SPNDs can be classified as prompt response and delayed response SPNDs. The emitter material can produce electrons directly following the neutron absorption reaction, such as in the case of rhodium (^{103}Rh) and vanadium (^{51}V) emitters. The disadvantage of using such

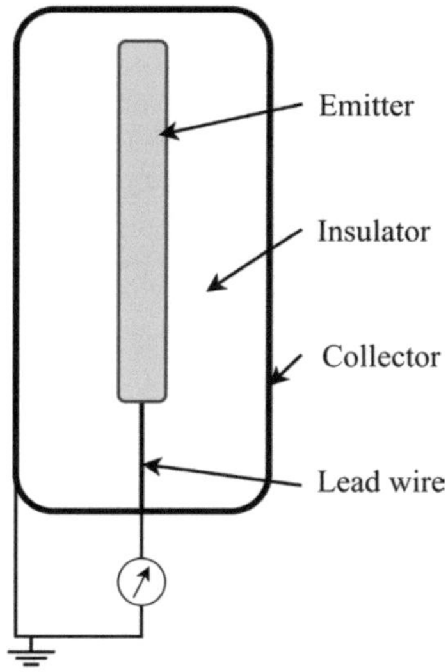

Figure 7.
Schematic drawing of a SPND.

emitters is that these materials decay with characteristic half-lives (0.7 and 3.76 minutes, respectively) [41]. Thus, the registered neutron flux is delayed; they are named delayed-response SPNDs.

Prompt SPNDs were designed to overcome this delay issue and obtain an instantaneous measurement of the neutron flux in the reactor core. Following the absorption of a neutron, the emitter nucleus will emit gamma rays which will interact with the emitter material to produce electrons. The time from the neutron absorption until the emission of electrons is extremely short. Thus, a prompt measurement of the neutron flux is obtained with such materials. Cobalt and cadmium are the main types of emitter materials in prompt SPNDs.

The selection of the emitter material for a SPND has been an important research topic. The half-life of the resulted radionuclide and thus the response required, the neutron absorption cross-section and therefore the burnup of the emitter material in the reactor core, the natural abundance of the isotope of interest and the ease in manufacturing it, and their adaptively to the high-temperature environment are the critical factors in the selection process.

From the above description of the main components of the SPND, it is clear that the SPND has several components that must be designed to operate in advanced reactors. One of the issues with the SPNDs is that they have a lifetime. This lifetime is affected by the degradation of the insulator material, to a lesser extent, the burnup of the emitter material with time. In addition, one of the issues is that the housing material should be compatible with the reactor coolant, which does not undergo corrosion.

The maximum working temperature for the SPNDs depends on the emitter material and the housing material properties. Typically, it is around 550°C [42]. SPNDs have not been used in high-temperature reactor environments or fast reactors (beyond light water reactor environments). However, deploying SPND in advanced reactors can improve the accuracy of power distribution measurements and hence the core operating LOCA and DNBR margin values [43]. The commercially available technology for LWRs is expected to have no problem in a high-pressure environment. However, a further study on the effects of the high-pressure environment on SPNDs is required.

The commercially available SPNDs cannot be used in fast reactors [44]. This problem can be explained by the drop in the absorption cross-section of the emitter material at higher neutron energies (absorption cross-section decreases with the increase in the incident neutron energy). For this reason, research is ongoing to select a proper material for the emitter to get good results in fast reactors [45, 46]. In molten salts, SPNDs require modifications in the housing material (exterior sheath) to be compatible with the salt [47].

Several funded projects by the NEET ASI program recently initiated the development of SPNDs for high-temperature environments. However, R&D projects are needed to modify the external housing material for the SPNDs if they are used in a chemically corrosive environment. In addition to that, the possibility of deployment of SPNDs in fast reactors is shown to be high. R&D programs can accelerate these efforts.

4.2 Gamma thermometers

Reactor power measurement sensors are based on detecting neutrons, gammas, or both. Gamma thermometers measure the reactor power by measuring the gamma-ray flux in the reactor core. In addition to that, gamma thermometers can be used to get information about core cooling in nearby channels. Furthermore, they can be used outside the reactor vessel.

The start of deployment of gamma thermometers goes back to the beginning of the 1950s in heavy water-moderated reactors. They were used because they are simple in design, very rugged, and insensitive to neutron fluence. They have the advantage over SPNDs that they do not deplete with time. Thus, they can operate longer than SPNDs and other in-core neutron detectors. Gamma thermometers are not suitable for startup or low power levels. At intermediate power level and up to full power, the gamma-ray flux is proportional to the reactor power level, and thus gamma thermometers can be used.

The design structure of gamma thermometers is similar to the thermocouples used in temperature measurement. A gamma thermometer is composed of a rod material, gap, and sheath [48]. The rod (inner body) is usually made of stainless steel. A gap filled with gas surrounding the inner body is used for heat insulation. The gas is selected with proper thermal conductivity, depending on the level of gamma-ray heating inside the core. For low levels of gamma-ray heating (< 2 W/g), xenon is a proper gap material. Argon is used for intermediate gamma-ray heating levels (~ 10 W/g), while helium is used for high levels of gamma-ray heating (~20 W/g) [49].

When inserted in the reactor core, incident gamma-ray flux heats the rod, which is transferred along the sensor axis. The gap filled with the gas will interrupt the heat flow. Two thermocouples measure the difference between the sheath and the steel rod regions. This difference is proportional to the gamma flux incident on the sheath and, thus, the power of nearby fuel rods. Several thermocouples can be used along the gamma thermometer to get the axial power distribution. Using gamma thermometers in several channels provides the radial power distribution. A simple schematic drawing of a gamma thermometer is shown in **Figure 8**.

Gamma thermometers have yet to achieve widespread, long-term deployment within U.S. commercial nuclear power plants [50]. In-plant tests as part of a joint research program were initiated to use gamma thermometers at commercial nuclear power plants in the United States and Japan [51]. In the United States, research supported by the NEET ASI program is ongoing to develop an optical fiber-based gamma thermometer [52]. Deployment of gamma thermometers in nuclear power plants can enhance safety and core monitoring while reducing other systems' operational and maintenance costs. In the Economic Simplified Boiling Water Reactor (ESBWR),

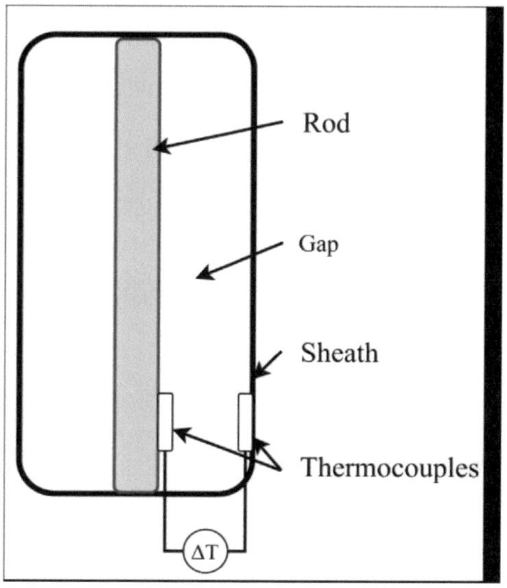

Figure 8.
Schematic drawing of a gamma thermometer.

gamma thermometers as in-core monitoring sensors are considered an essential feature over the previous BWRs [53].

Gamma thermometers have not been demonstrated at the higher temperatures of VHTRs. A significant redesign is needed to function at higher temperatures [54]. However, with their simplicity in design, it is promising that they would perform well at higher temperatures. For MSRs, customization effort would be required to develop a version suitable for the reactor environment. High pressure can exert mechanical stress on the components of gamma thermometers. This stress may cause deformation, cracking, or failure of structural elements. Thus, this high pressure may reduce the lifetime of such components. R&D programs must be initiated to study the effect of a high-pressure environment on gamma thermometers.

4.3 Fission chambers

Fission chambers are a type of ionization chamber. Unlike SPNDs, fission chambers require the application of an electric voltage between the cathode and the anode. The chamber is filled with a gas (commonly argon) at high pressure. Fission chambers can be used as in-core or ex-core nuclear instrumentation. A schematic drawing of a fission chamber is shown in **Figure 9**.

By placing a fission chamber in a neutron field, incident neutrons will be absorbed by the fissile material lined on the fission chamber housing. Usually, the housing material is steel or aluminum, while the fissile material is uranium highly enriched in a ^{235}U isotope. Using highly enriched uranium increases the fission reaction rate and thus enhances the ionization current.

Upon ^{235}U fission, two ionized fission products will be produced with a total kinetic energy of around 160 MeV. These two fission fragments will travel in opposite directions. One of the two fission fragments will be traveling into the gas chamber. For this reason, the thickness of the fissile material should be optimized to allow the fission fragment to reach the gas without affecting the fission rate [55]. The fission

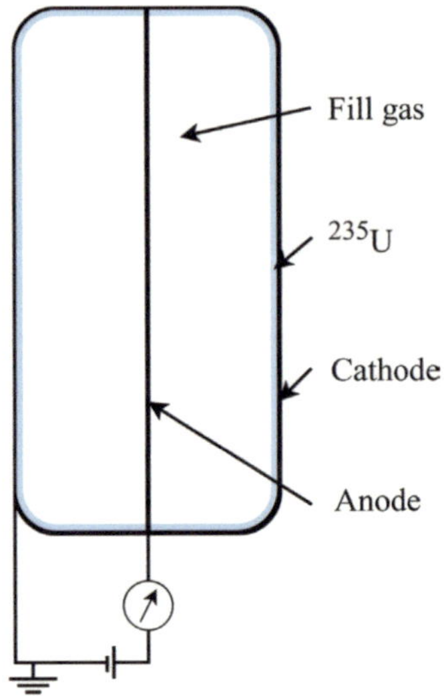

Figure 9.
Schematic drawing of a fission chamber.

fragment will ionize the filled gas atoms along its path, producing positive ions and negative electrons. An external voltage is applied to the cathode and anode, causing the positive ions to move toward the cathode while the electrons are collected on the anode. The current produced from this process is proportional to the fission rate in the fissile material and, thus, to the reactor core's neutron flux [56].

The depletion of the fissile material is the leading cause of the limited operating time for fission chambers as in-core nuclear instrumentation. The sensitivity of a fission chamber decreases with time because of the neutron fluence. Combining fissile and fertile materials can be a solution to such a problem.

Commercial fission chambers are available for the operating environment of LWRs. However, no commercial fission chamber is available for high-temperature reactor environments (above 550°C). Locating the chambers ex-core will reduce the temperature to those that commercially available fission chambers can withstand. The housing material and the internal gas pressure determine the operating temperatures of fission chambers [57].

High pressure may affect the sealing and containment of a fission chamber. The fission chamber must be appropriately sealed to prevent pressure leakage, which could compromise its functionality. Specialized sealing materials and techniques capable of withstanding high-pressure differentials may be required. There is no available research on the effect of a high-pressure environment on fission chambers.

A fission chamber that would operate at high temperatures of 550°C and beyond is named High-Temperature Fission Chamber (HTFC). Previous research programs have developed such chambers that run up to 800°C. However, these chambers are not commercially available. Developing HTFCs and demonstrating their sensitivity and mechanical and thermal robustness in the high-temperature environment will pave the way for developing high-temperature reactors. Several DOE NEET ASI projects have been initiated to design, fabricate, and demonstrate High-Temperature Fission Chambers (HTFCs). They aim to provide HTFC for HTRs, MSRs, and fast reactors.

Micro-pocket fission chambers are pancake-style, highly miniaturized fission chambers that employ sealed alumina plates as their structural backings and coatings of uranium or thorium as their neutron-sensitive element [58]. Micro-pocket fission chambers are a promising technology used in a high-temperature environment. The technology has yet to be commercially available. Several funded projects by DOE NEET ASI were initiated to develop it.

4.4 Ionization chambers

In PWRs, the core power density distribution can be measured using ex-core detectors. These detectors can be installed outside the reactor vessel in several axial locations. The measured neutron flux in these detectors is then used to get the axial power distribution in the reactor core.

Ionization chambers are neutron detectors that can be used outside the reactor vessel for neutron flux measurements. They are gas-filled radiation detectors that can be designed to measure several types of ionizing radiation. The principle of operation is similar to fission chambers. Compared to fission chambers, ionization chambers use ^{3}He or ^{10}B nuclides as the filling gas instead of ^{235}U to produce the charged particles.

Compensated and uncompensated ionization chambers are used as ex-core nuclear instrumentation. Uncompensated ionization chambers are used as a power-range monitoring system. Compensated ionization chambers are used for intermediate power range. The compensated ionization chamber comprises two distinct chambers: one is coated with boron, while the other is uncoated. The coated chamber exhibits sensitivity to gamma and neutrons, whereas the uncoated chamber only detects gamma rays. The detector output is the difference between the two currents generated from the chambers. This difference in current represents the neutron response of the compensated ionization chamber. Therefore, the compensated ionization chamber cancels the current resulting from the gamma rays' interaction. This compensation technique proves valuable in the intermediate power range, ensuring that the detector response is only due to neutron flux.

Several types of compensated and uncompensated ionization chambers are used in LWRs and high-temperature reactors. Thus, no significant challenge is expected in deploying the currently available technology in the advanced high-temperature reactor environment. In a fast reactor environment, ex-core moderation is required to achieve better sensitivities in the measurements or to use more giant ionization chambers.

In MSRs, ionization chambers are expected to be compatible and thus have no problem deploying such detectors [59]. This compatibility can be explained by the fact that these detectors will be used as ex-core nuclear instrumentation. In fact, in the only two MSRs built, no neutron instrumentation was employed in the core. Only ex-core detectors were used. However, for the proposed advanced MSRs, employing in-core neutron detectors that can withstand their harsh environment is necessary for high-sensitivity measurements.

4.5 Proportional counters

Proportional counters are gas-filled detectors used as a source-range neutron flux monitoring system. They are used as ex-core nuclear instrumentation during the first criticality or for start-up after a very long shutdown. Their principal mode of operation is quite similar to the ionization chamber. However, the applied voltage to the cathode and anode is higher than in ionization chambers.

Some proportional counters are based on the (n, α) reaction with ^{10}B in the boron trifluoride gas (BF_3) filled in the detector's chamber. Boron-lined proportional counters have been used in High-Temperature Engineering Test Reactor (HTTR) for temperatures up to 600°C. In gas-cooled reactors, proportional counters have been used as neutron sensors [60]. Thus, no temperature problem is expected to arise in high-temperature reactor environments. Therefore, commercially available proportional counters technology is mature enough, and R&D is not a priority for such a type of detector.

Proportional counters can be used to detect thermal neutrons or fast neutrons. For fast-spectrum reactors, proportional counters are less efficient due to the ^{10}B capture cross-section drop with increased neutron energy. Using more giant detectors can solve this problem to compensate for this drop in cross section. In addition to that, helium or a gas with a low atomic number can be used in the chamber to moderate the neutrons. In molten salt reactors, proportional counters are expected to operate with no problem since they are deployed out of the corrosive environment of the reactor core.

4.6 On-demand nuclear instrumentation

Some nuclear instrumentation systems are on-demand operation. That is, the system is not in a continuous mode of operation and may not be installed in fixed core positions [61]. One of the examples of such a system is the Aeroball Measurement System (AMS) used in the Evolutionary Power Reactor (EPR). The system measures the neutron flux in predetermined three-dimensional positions in the reactor core. In addition to that, AMS is used to calibrate SPNDs and predict power density distribution in the reactor core [62].

The AMS is composed of several steel balls that contain vanadium. These balls move in specific vertical tubes in the reactor core by a nitrogen gas driving mechanism. The aeroballs enter the reactor core using nitrogen gas pressure. In the reactor core, the thermal neutron flux will irradiate the aeroballs, and many of the ^{51}V atoms will capture neutrons resulting in ^{52}V in an excited state with a half-life of 3.74 minutes. ^{52}V will undergo beta decay with the emission of about 1.43 MeV gamma-ray.

After irradiation, the aeroballs are transported to waiting positions before they are transported to a measuring system. The measuring system consists of several planar silicon detectors that measure the gamma-ray intensity resulting from the decay of the activated atoms. Information about the relative neutron flux values in the reactor core and, thus, power distribution can be calculated using software using these measured intensities from the aeroballs [63]. Because of the short half-life of the produced ^{52}V, this measurement process can be repeated every 15 minutes allowing for the reuse of the aeroballs [64].

Even though such on-demand nuclear instrumentation systems may be installed in the reactor core, they will face a different reactor operation environment than other in-core nuclear sensors and instrumentation. This results from the fact that they will not be exposed to the harsh in-core environment for the same duration. Therefore, they may be treated like ex-core nuclear sensors and instrumentation. In addition, unlike fission chambers or gamma thermometers, such systems are mainly used in certain reactors and not widely used in other reactor designs. For these reasons, the on-demand nuclear instrumentation systems are not discussed further.

5. Overall assessment

The compatibility of the nuclear sensors and instrumentation with the advanced reactor environments was assessed. For in-core nuclear instrumentation, it is clear that some of the nuclear sensors and instrumentation technologies used in current LWRs may not be suitable for the advanced reactor environment. SPNDs are expected to have issues operating in a fast-spectrum environment, high temperature, and corrosive environment. Gamma thermometers are not ready to be used in high-temperature or corrosive environments. Fission chambers are compatible with the fast spectrum environment and high-temperature environment. However, the technology is not commercially available at high temperatures. The housing material must be modified to operate in a corrosive environment.

Overall, it can be demonstrated that the ex-core nuclear instrumentation is expected to have no issue when deployed in advanced nuclear reactors. This fact is mainly because, outside the reactor core, the neutron flux, temperature, and corrosion environment are several folds lower than that in the reactor core. Thus, the ex-core environment in the advanced reactors will be the same as in the currently operating reactors. Using ex-core nuclear instrumentation may be sufficient for reactor operation and safety. On the other hand, deploying in-core nuclear sensors and instrumentation may increase the sensitivity of the measurements and, thus, the overall safety. A cost/benefit analysis is required to assess the need to deploy the nuclear sensors and instrumentation in the in-core advanced reactor environments to decide whether an R&D is necessary to deploy the technology.

The compatibility of the above-discussed in-core and ex-core nuclear sensors and instrumentation with various reactor environments is summarized in **Tables 3** and **4**, respectively. Since LWRs have a long operating experience, the comparison was performed with the environment in LWRs.

Reactor Environment	In-core nuclear instrumentation		
	SPNDs	**Gamma thermometers**	**Fission chambers**
LWR	**Compatible** They have been used in thermal nuclear reactors for more than 60 years.	**Compatible** Have been used since the 1950s.	**Compatible** Have been widely used.
Fast spectrum	**Incompatible** The emitter material should be replaced.	**Compatible** Insensitive to neutron fluence.	**Compatible** Larger size detectors are used.
High temperature	**Incompatible** A maximum working temperature is around 550°C.	**Incompatible** A significant redesign is required to function at higher temperatures.	**Compatible** These chambers are not commercially available.
Molten salts	**Incompatible** The housing material should be modified for salt compatibility.	**Incompatible** A version suitable for the reactor environment is required.	**Incompatible** The housing material should be modified for salt compatibility.
High pressure	**Compatible** The effect of high pressure on its lifetime and structural integrity should be studied.	**Compatible** The effect of high pressure on its lifetime and structural integrity should be studied.	**Compatible** The effect of high pressure on its sealing and containment should be studied.

Table 3.
Compatibility of the in-core nuclear instrumentation with various reactor environments.

Reactor Environment	Ex-core nuclear instrumentation		
	Fission chambers	**Ionization Chambers**	**Proportional Counters**
LWR	**Compatible** Several types are being used worldwide.	**Compatible** Compensated and uncompensated types are being widely used.	**Compatible** They are being used.
Fast spectrum	**Compatible** Larger size detectors are usually used.	**Compatible** Larger size detectors are usually used.	**Compatible** Larger size detectors are usually used.
High temperature	**Compatible** Ex-core reactor environment would be the same as in LWRs.	**Compatible** They have been used in several gas-cooled reactors.	**Compatible** The commercially available counters can be used.
Molten salts	**Compatible** They were used as ex-core nuclear instrumentation.	**Compatible** They were used as ex-core nuclear instrumentation.	**Compatible** They were used as ex-core nuclear instrumentation.
High pressure	**Compatible** Ex-core reactor environment would be the same as in LWRs.	**Compatible** Ex-core reactor environment would be the same as in LWRs.	**Compatible** Ex-core reactor environment would be the same as in LWRs.

Table 4.
Compatibility of the ex-core nuclear instrumentation with various reactor environments.

Although in-core nuclear instrumentation is expected to function at high pressure, the high-pressure environment may impact the in-core nuclear instrumentation. This influence can be by affecting the structural integrity of the instrumentation caused by the mechanical stresses applied by the high pressure. This stress may cause deformation, cracking, or failure of structural elements of the nuclear instrumentation over the operating time. Additionally, the high pressure may affect the sealing and containment of instrumentation filled with gasses. Thus, instruments must be appropriately sealed to prevent pressure leakage, which could compromise their functionality. Furthermore, the high pressure may influence the calibration standards used for pressure-sensitive instruments. For these reasons, the compatibility and durability of nuclear sensors and instrumentation with high-pressure reactor environments such as SCWR must be investigated.

Table 5 summarizes the technology gaps, commercial availability, and the R&D needs for the above-discussed in-core nuclear instrumentation in currently operating PWRs, BWRs, and generation IV reactors.

To assess the priorities in future research projects, the availability of funded research projects by the DOE NEET ASI program in advanced nuclear sensors and instrumentation was investigated [65–70]. An overview of the DOE NEET ASI program-funded projects on nuclear sensors and instrumentation is listed in **Table 6**.

It can be observed that all nuclear sensors and instrumentation-funded projects were in-core sensors and instrumentation. This trend supports the conclusions of the study that R&D priorities are for the in-core nuclear reactor sensors and instrumentation. It is shown that nearly all research projects were focused on the high-temperature environment. Thus, R&D projects are required to study in-core nuclear instrumentation in fast-spectrum, high-pressure, and corrosive environments.

Reactor Environment	In-core nuclear instrumentation		
	SPNDs	Gamma thermometers	Fission chambers
PWR	**The technology is mature.** Commercially available.	**The technology is mature.** Have been widely used and is commercially available.	**The technology is mature.** Have been widely used and is commercially available.
BWR	**The technology is mature.** Commercially available.	**The technology is mature.** Have been widely used and is commercially available.	**The technology is mature.** Have been widely used and is commercially available.
SCWR	**Requires R&D** The commercially available technology is expected to function well in a high-pressure environment. However, the durability and structural integrity may need to be studied in a high-pressure environment.		
VHTR	**Requires R&D** to operate at high temperatures. Not commercially available.	**Requires R&D** to operate at high temperatures. Not commercially available.	**Requires R&D** to operate at high temperatures. Not commercially available.
SFR	**Requires R&D** to operate at high temperatures with the fast spectrum environment. Not commercially available.	**Requires R&D** Not commercially available.	**Requires R&D** to operate at high temperatures. Not commercially available.
LFR	**Requires R&D** to operate in the fast spectrum environment.	**R&D is not a priority** Commercially available technology can be used.	**R&D is not a priority** Commercially available technology can be used.
GFR	**R&D is required** to operate in a fast-spectrum environment.	**Requires R&D** to operate at high temperatures. Not commercially available.	**Requires R&D** to operate at high temperatures. Not commercially available.
MSR	**R&D is required** to modify the emitter and housing materials and to operate in a fast-spectrum environment. Not commercially available.	**R&D is required** to function at high temperatures and to be salt compatible. Not commercially available.	**Requires R&D** to operate at high temperatures. Housing material should be modified. Not commercially available.

Table 5.
Summary of in-core nuclear instrumentation R&D needs in current and advanced reactors.

6. Summary

Advanced nuclear reactors (Generation IV) will have differences in operating environments from the currently operating nuclear reactors. These differences are mainly the higher operating temperature and pressure, the corrosive environment, and the fast neutron spectrum. Developing sensors and instrumentation systems capable of operating in these advanced reactors will play a significant role in their licensing and demonstration efforts.

Nuclear instrumentation	Project Description	Project year(s)
Self-Powered Neutron Detector	**Nuclear Instrumentation - Neutron Flux Sensors** Demonstration and qualification of Self-Powered Neutron Detectors (SPNDs).	2020
	Flux Measurement Calibration processes were developed, focusing on their performance at high temperatures.	2020–2021
	Irradiation Testing of Neutron Flux Sensors in the Advanced Test Reactor Critical Facility The aim is to test neutron flux in a variety of reactor environments.	2020–2021
Fission Chambers	**Micro Pocket Fission Detectors** Development of Micro Pocket Fission Detectors.	2012–2014
	High-Temperature Fission Chamber Fabricate and test a high-temperature fission chamber for temperatures up to 800°C.	2012–2014
	Enhanced Micro-Pocket Fission Detector for High-Temperature Reactors Demonstrate and deploy a high-temperature micro-pocket fission chamber for temperatures up to 800°C.	2014–2017
	Flux Measurement Calibration processes are being developed, focusing on their performance at high temperatures.	2020–2021
Gamma Thermometers	**Development of an Optical Fiber-Based Gamma Thermometer** Build an optical fiber-based gamma thermometer (OFBGT) and test it in two research reactors.	2018–2021

Table 6.
DOE NEET ASI program-funded research projects.

It was demonstrated that the in-core reactor sensors and instrumentation are prioritized in future R&D projects. R&D is required mainly for SPNDs, gamma thermometers, and fission chambers. SPNDs have to be able to operate at higher temperatures, with a fast neutron spectrum, and in a molten salt reactor environment. Gamma thermometers require R&D projects to work in a fast neutron spectrum reactor and with a molten salt environment. Fission chambers that would be employed as in-core reactor instrumentation require R&D to commercialize the available technology of fission chambers to operate in a high-temperature reactor environment. In addition to that, they need R&D if they will be used in a molten salt environment. The durability and structural integrity of all these in-core instrumentation has to be further investigated in a high-pressure environment.

Ex-core nuclear instrumentation is expected to face the same operational environments as those in currently operating LWRs. Thus, R&D for ex-core nuclear instrumentation is not a priority. On-demand in-core nuclear instrumentation must be evaluated based on operating time in the reactor environment. In addition, some types of nuclear instrumentation are not used in the in-core reactor environment. Employment of such sensors and instrumentation in advanced reactors may increase the accuracy of the measurement and, thus, the reactor's safety. However, cost/benefit analysis is necessary for such R&D projects.

In the United States, the Department of Energy established many research programs to fill the technological gaps and assess the advanced reactor needs of a new generation of sensors and instrumentation. Many R&D projects have been initiated and supported by the NEET ASI program in the last few years. For this reason, where R&D is required based on the defined approach, the availability of initiated research projects on the sensor and instrumentation technology was further investigated. It was shown that the funded projects in the last few years were focused on in-core nuclear sensors and instrumentation. However, R&D projects for gamma thermometers at high temperatures are required. In addition, if SPNDs are planned to be used in MSRs, then R&D projects are necessary to design SPNDs compatible with the corrosive environment.

The assessment guides the most prior R&D needs in nuclear sensors and instrumentation technologies. Filling the gaps in the technology will enable advanced reactors to follow the proven safe and operational path of previously operated nuclear reactors.

Author details

Thabit Abuquidaira[1], Pavel Tsvetkov[2]* and Piyush Sabharwall[3]

1 Texas A&M University, USA

2 Department of Nuclear Engineering, Texas A&M University, College Station, Texas, United States

3 Idaho National Laboratory, Idaho, USA

*Address all correspondence to: tsvetkov@tamu.edu

References

[1] International Atomic Energy Agency. Power Reactor Information System. Vienna, Austria: International Atomic Energy Agency. 2022. Available from: https://pris.iaea.org/pris/ [Accessed: October 18, 2022]

[2] Generation IV International Forum. When Will Gen IV Reactors Be Built? Paris, France: Generation IV International Forum; 2022. Available from: https://www.gen-4.org/gif/jcms/c_41890/faq-2 [Accessed: October 19, 2022]

[3] International Atomic Energy Agency. Nuclear Power Plant Instrumentation and Control: A Guidebook. Vienna, Austria: International Atomic Energy Agency; 1984

[4] Meininger R. Three Mile Island Technical Information and Examination Program Instrumentation and Electrical Summary Report. Washington D.C., United States: Department of Energy; 1985

[5] Rempe J, Knudson Darrell. TMI-2—A Case Study for PWR Instrumentation Performance during a Severe Accident [Online]. 2013. Available from: http://www.inl.gov

[6] Nuclear Regulatory Commission. Report to Congress: Advanced Reactor Licensing. Washington D.C., United States: United States Nuclear Regulatory Commission; 2012

[7] U.S. Department of Energy-Office of Nuclear Energy. Light Water Reactor Sustainability (LWRS) Program. Washington D.C., United States: U.S. Department of Energy-Office of Nuclear Energy; 2022. Available from: https://www.energy.gov/ne/nuclear-reactor-technologies/light-water-reactor-sustainability-lwrs-program [Accessed: October 27, 2022]

[8] U.S. Department of Energy-Office of Nuclear Energy. Nuclear Energy Enabling Technologies (NEET). Washington D.C., United States: U.S. Department of Energy-Office of Nuclear Energy; 2022. Available from: https://www.energy.gov/ne/nuclear-energy-enabling-technologies-neet [Accessed: July 24, 2022]

[9] Locatelli G, Mancini M, Todeschini N. Generation IV nuclear reactors: Current status and future prospects. Energy Policy. 2013;**61**:1503-1520. DOI: 10.1016/j.enpol.2013.06.101

[10] Lamarsh J, Baratta A. Introduction to Nuclear Engineering. 3rd ed. Addison-Wesley. Upper Saddle River, New Jersey: Prentice Hall; 2001

[11] United States Nuclear Regulatory Commission. Power Reactors. Washington D.C., United States: United States Nuclear Regulatory Commission; 2022. Available from: https://www.nrc.gov/reactors/power.html [Accessed: October 23, 2022]

[12] Rippon S. History of the PWR and its worldwide development. Energy Policy. 1984;**12**(3):259-265

[13] Subki MH. Water cooled small modular reactors (integral PWR and BWR). In: Encyclopedia of Nuclear Energy. New York, United States: Elsevier; 2021. pp. 694-710. DOI: 10.1016/b978-0-12-819725-7.00208-7

[14] Westinghouse. International Reactor Innovative and Secure (IRIS) Plant Overview. Washington D.C., United States: Westinghouse; Nuclear regulatory Commission, 2002

[15] Ray HB. U.S. options for licensing a new commercial power plant. In: Encyclopedia of Nuclear Energy. New York, United States: Elsevier; 2021. pp. 168-174. DOI: 10.1016/b978-0-12-409548-9.12159-1

[16] Hamon DA. Boiling water reactors. In: Encyclopedia of Nuclear Energy. New York, United States: Elsevier; 2021. pp. 214-235. DOI: 10.1016/B978-0-12-819725-7.00027-1

[17] Wu P, Ren Y, Feng M, Shan J, Huang Y, Yang W. A review of existing SuperCritical water reactor concepts, safety analysis codes and safety characteristics. In: Progress in Nuclear Energy. Vol. 153. New York, United States: Elsevier Ltd; 2022. DOI: 10.1016/j.pnucene.2022.104409

[18] Cai J, Renault C, Gou J. Supercritical water-cooled reactors. Science and Technology of Nuclear Installations. 2014;**2014**:548672. DOI: 10.1155/2014/548672

[19] Wang J, Wang Q, Ding M. Review on Neutronic/thermal-hydraulic coupling simulation methods for nuclear reactor analysis. Annals of Nuclear Energy. 2020;**137**:107165. DOI: 10.1016/j.anucene.2019.107165

[20] Rahman MM, Dongxu J, Jahan N, Salvatores M, Zhao J. Design concepts of supercritical water-cooled reactor (SCWR) and nuclear marine vessel: A review. In: Progress in Nuclear Energy. Vol. 124. New York, United States: Elsevier Ltd; 2020. DOI: 10.1016/j.pnucene.2020.103320

[21] Tsvetkov PV, Lewis TG, Alajo AB, Ii DEA. VHTR-based systems for autonomous co-generation applications. Nuclear Engineering and Design. 2010;**240**:2451-2457. DOI: 10.1016/j.nucengdes.2010.05.053

[22] Scherr J, Tsvetkov P. Reactor design strategy to support spectral variability within a sodium-cooled fast spectrum materials testing reactor. Annals of Nuclear Energy. 2018;**113**:15-24. DOI: 10.1016/j.anucene.2017.10.049

[23] Abram T, Ion S. Generation-IV nuclear power: A review of the state of the science. Energy Policy. 2008;**36**(12):4323-4330. DOI: 10.1016/j.enpol.2008.09.059

[24] Hatala B. Gas cooled fast reactor system (GFR). In: Encyclopedia of Nuclear Energy. New York, United States: Elsevier; 2021. pp. 545-552. DOI: 10.1016/b978-0-12-409548-9.12207-9

[25] Tsvetkov P. Gas-cooled fast reactors (GFRs). In: Handbook of Generation IV Nuclear Reactors. New York, United States: Woodhead Publishing, Elsevier; 2023. pp. 167-172. DOI: 10.1016/B978-0-12-820588-4.00016-5

[26] Abuqudaira TM, Stogov YV. Possibilities of better utilization of MOX fuel in VVER type reactors by optimizing neutron spectrum. Journal of Physics: Conference Series. 2020;**1689**:012023. DOI: 10.1088/1742-6596/1689/1/012023

[27] Bettis ES, Schroeder RW, Cristy GA, Savage HW, Affel RG, Hemphill LF. The aircraft reactor experiment -design and construction. Nuclear Science and Engineering. 1957;**2**(6):804-825. DOI: 10.13182/nse57-a35495

[28] Prince BE, Ball SJ, Engel JR, Haubenreich PN, Kerlin TW. Zero-Power Physics Experiments on the Molten-Salt Reactor Experiment. ORNL-4233; Oak Ridge, Tennessee: Oak Ridge National Laboratory; 1968

[29] Serp J et al. The molten salt reactor (MSR) in generation IV: Overview and perspectives. Progress in Nuclear

Energy. 2014;**77**:308-319. DOI: 10.1016/j.pnucene.2014.02.014

[30] NSA. Nuclear Suppliers Association. Available from: https://nuclearsuppliers.org/ [Accessed: July 19, 2022].

[31] NASA. Technology Readiness Level. Available from: https://www.nasa.gov/directorates/heo/scan/engineering/technology/technology_readiness_level [Accessed: July 19, 2022]

[32] Hashemian HM. Nuclear power plant instrumentation and control. In: Tsvetkov P, editor. Nuclear Power - Control, Reliability and Human Factors. London, United Kingdom: IntechOpen; 2011

[33] Knoll G. Radiation Detection and Measurement. 4th ed. Hoboken, New Jersey, United States: Wiley; 2010

[34] Westinghouse Electric Company LLC. Chapter 7: Instrumentation and Controls, AP1000 Design Control Document. Westinghouse Electric Company LLC. Washington D.C., United States: Nuclear Regulatory Commission; 2011

[35] Pacific Northwest National Laboratory. Technical Readiness and Gaps Analysis of Commercial Optical Materials and Measurement Systems for Advanced Small Modular Reactors - PNNL-22622, Rev. 1. Richland, Washington: Pacific Northwest National Laboratory; 2013

[36] U.S.NRC. AP1000 Design Control Document, Chapter 4. Washington D.C., United States: U.S.NRC, Nuclear Regulatory Commission; 2011

[37] Sang Y et al. Development and verification of a simulation toolkit for self-powered neutron detector. Annals of Nuclear Energy. 2021;**150**:107784. DOI: 10.1016/j.anucene.2020.107784

[38] Liu X, Wang Z, Zhang Q, Deng B, Niu Y. Current compensation for material consumption of cobalt self-powered neutron detector. Nuclear Engineering and Technology. 2020;**52**(4):863-868. DOI: 10.1016/j.net.2019.09.010

[39] Harrer JM, Beckerley JG. Nuclear Power Reactor Instrumentation Systems Handbook. Vol. 11973. Washington D.C., United States: U.S. Atomic Energy Commission, Nuclear Regulatory Commission; 1974

[40] Cui T, Yang Y, Xue H, Kuang H. A Monte-Carlo simulation method for the study of self-powered neutron detectors. In: Nuclear Instruments and Methods in Physics Research, Section a: Accelerators, Spectrometers, Detectors and Associated Equipment. Vol. 954. New York, United States: Elsevier B.V; 2020. DOI: 10.1016/j.nima.2018.10.061

[41] U.S. Department of Energy. DOE Fundamentals Handbook Instrumentation and Control Volume 2 of 2. Vol. 2. Washington D.C., United States: U.S. Department of Energy; 1992

[42] Goodings A. Experience with high-temperature radiation detectors and cables for reactor instrumentation systems. In: Nuclear Power Plant Control and Instrumentation, Proceedings of a Symposium. Vienna: International Atomic Energy Agency; 1978. pp. 225-242

[43] Mourlevat J, Janvier D, Warren H. Industrial tests of rhodium self-powered detectors: The Golfech 2 experimentation. In: Core Monitoring for Commercial Reactors: Improvements in Systems and Methods. Workshop Proceedings, Stockholm, Sweden, October 4-5, 1999. Paris, France: OEECD NEA; 2000. pp. 65-77

[44] Goetz KC, Cetiner SM, Celik C. Development of a fast-Spectrum

self-powered neutron detector for molten salt experiments in the versatile test reactor. EPJ Web of Conferences. 2021;**253**:05006. DOI: 10.1051/epjconf/202125305006

[45] Verma V, Barbot L, Filliatre P, Hellesen C, Jammes C, Svärd SJ. Self powered neutron detectors as in-core detectors for sodium-cooled fast reactors. Nuclear Instruments and Methods in Physics Research. 2017;**860**:6-12. DOI: 10.1016/j.nima.2017.04.011

[46] Angelone M, Klix A, Pillon M, Batistoni P, Fischer U, Santagata A. Development of self-powered neutron detectors for neutron flux monitoring in HCLL and HCPB ITER-TBM. Fusion Engineering and Design. 2014;**89**(9-10):2194-2198. DOI: 10.1016/j.fusengdes.2014.01.077

[47] Holcomb DE, Kisner RA, Cetiner SM. Instrumentation Framework for Molten Salt Reactors [online]. 2018. Available from: www.osti.gov

[48] Birri A, Blue TE. Methodology for inferring reactor core power distribution from an optical fiber based gamma thermometer array. Progress in Nuclear Energy. 2020;**130**:103552. DOI: 10.1016/j.pnucene.2020.103552

[49] Van Nieuwenhove R, Vermeeren L. Nuclear heating measurements by gamma and neutron thermometers. EPJ Web of Conferences. 2020;**225**:04003. DOI: 10.1051/epjconf/202022504003

[50] Nuclear Regulatory Commission. Instrumentation in VHTRs for Process Heat Applications. Washington D.C., United States: Nuclear Regulatory Commission; 2010

[51] Raghavan R, Martin CL, Wirth AL, Itoh T, Goto Y, Arai R. Application of the gamma thermometer as BWR fixed In-Core calibration system. In: Proceedings of the Specialists Meeting on Incore Instrumentation and Reactor Core Assessment, Mito City, Japan. Paris, France; 1996

[52] Birri A, Petrie CM, Blue TE. Parametric analysis of an optical fiber–based gamma thermometer for university research reactors using an analytic thermal model. Nuclear Technology. 2021;**207**(12):1865-1872. DOI: 10.1080/00295450.2020.1844532

[53] Theriault K. Boiling water reactors (BWRs). In: Kok K, editor. Nuclear Engineering Handbook. Boca Raton, Florida, United States: CRC Press; 2009. pp. 83-139

[54] Oak Ridge National Laboratory. HTGR Measurements and Instrumentation Systems-ORNL/TM-2012/107 [Online]. Oak Ridge, Tennessee, United States: Oak Ridge National Laboratory; 2012. Available from: http://www.osti.gov/contact.html

[55] Reilly D, Ensslin N, Smith H, Kreiner S, Unis E, Los Alamos National Laboratory (U.S.). Nuclear Regulatory Commission. Neutron detectors. In: Passive Nondestructive Assay of Nuclear Materials, US Department of Commerce. Washington D.C., United States: National Technical Information Service; 1991. pp. 379-406

[56] Coburn J, Luker SM, Parma EJ, DePriest KR. Modeling, calibration, and verification of a fission chamber for ACRR experimenters. In: EPJ Web of Conferences, EDP Sciences. Paris, France; 2016. DOI: 10.1051/epjconf/201610605001

[57] Lamphere R. Fission detectors. In: Marion J, Fowler J, editors. Fast Neutron Physics. Part 1. Techniques. New York: Interscience; 1960

[58] McGregor DS, Ohmes MF, Ortiz RE, Sabbir Ahmed ASM, Kenneth Shultis J. Micro-pocket fission detectors (MPFD) for in-core neutron flux monitoring. Nuclear Instruments and Methods in Physics Research A. 2005;**554**(1-3):494-499. DOI: 10.1016/j.nima.2005.06.086

[59] Oak Ridge National Laboratory. Instrumentation Framework for Molten Salt Reactors - ORNL/TM-2018/868 [Online]. Oak Ridge, Tennessee, United States: Oak Ridge National Laboratory; 2018. Available from: https:// www.osti.gov

[60] Oak Ridge National Laboratory. Assessment of Sensor Technologies for Advanced Reactors - ORNL/TM-2016/337 R1 [Online]. Oak Ridge, Tennessee, United States: Oak Ridge National Laboratory; 2016. Available from: http://www.osti.gov/scitech/

[61] AREVA NP. U.S. EPR Nuclear Incore Instrumentation Systems Report. Paris, France: AREVA NP; 2006

[62] Dias AM, Silva FC. Determination of the power density distribution in a PWR reactor based on neutron flux measurements at fixed reactor incore detectors. Annals of Nuclear Energy. 2016;**90**:148-156. DOI: 10.1016/j.anucene.2015.12.002

[63] Glasow PA. Aeroball system and energy-dispersive analysis: Important industrial applications of silicon detectors. Nuclear Instruments and Methods in Physics Research. 1984;**226**:17-25

[64] Konheiser J, Mueller SE, Seidl M. Study of the influence of water gaps between fuel assemblies on the activation of an aeroball measurement system (AMS). Annals of Nuclear Energy. 2020;**136**:107005. DOI: 10.1016/j.anucene.2019.107005

[65] U.S. Department of Energy. Advanced Sensors and Instrumentation Project Summaries. U.S. Department of Energy; 2016

[66] U.S. Department of Energy. Advanced Sensors and Instrumentation Project Summaries. Washington D.C., United States: U.S. Department of Energy; 2017

[67] U.S. Department of Energy. Advanced Sensors and Instrumentation Project Summaries. Washington D.C., United States: U.S. Department of Energy; 2018

[68] U.S. Department of Energy. Advanced Sensors and Instrumentation Project Summaries. Washington D.C., United States: U.S. Department of Energy; 2019

[69] U.S. Department of Energy. Advanced Sensors and Instrumentation Project Summaries. Washington D.C., United States: U.S. Department of Energy; 2020

[70] U.S. Department of Energy. Advanced Sensors and Instrumentation Project Summaries. Washington D.C., United States: U.S. Department of Energy; 2021

Chapter 5

Perspective Chapter: PRA and Protective System Maintenance

Ernie Kee and Martin Wortman

Abstract

The processes used to manage protective system equipment failures as they relate to Probabilistic Risk Assessment (PRA) in the commercial nuclear power setting are reviewed. Efficacy of protection is governed by maintenance policy that includes system modification, maintenance inter-arrivals as a function of time, and upset inter-arrivals as a function of time. Such a maintenance policy is the one used in nuclear power plant protective systems. Observations described in this article include the impact of time-dependent activities associated with maintenance policy as they relate to endogenous and exogenous upset inter-arrival times. Methods evaluating maintenance policy reliant on combinatorial logic, such as PRA, fault trees, or event trees, may lead to ineffective maintenance policy decision-making for protective system efficacy. Recommendations for maintaining effective protections, and connections to engineering maintenance practice and regulations are made based on the implications that come from our observations. The importance of the issues described herein is that the relationship of design, maintenance, and repair policies must be properly understood and taken into account by process owners, operators, and investors, as well as regulators who specify and enforce protections in hazardous processes.

Keywords: PRA, protection, protective systems, maintenance efficacy, nuclear power

1. Introduction

On March 28, 1979 the United States commercial nuclear power program experienced its first major accident; the reactor core in Unit 2 at the Three Mile Island nuclear plant site in Harrisburg, Pennsylvania had overheated and melted down. Even considering this accident, it can be said, based on the safety record of commercial nuclear power in the United States, that the Nuclear Regulatory Commission (NRC) has successfully produced a regulatory structure that effectively manages risk of radioactive releases, especially from commercial nuclear power plants.[1] With the benefit of hindsight, this accident would help inform stakeholders to better manage risk from nuclear power plant accidents. Within the nuclear power industry, the Three Mile Island accident motivated development of the predictive modeling

---

[1] The President's Commission report "The Accident at Three Mile Island" states that the level of radioactivity released from the Three Mile Island accident was small Kemeny (see [1], p. 34).

methods and risk analytics commonly in use today. Thus, before exploring the details of reactor protection risk analytics, it is important to review key engineering insights gained following the Three Mile Island accident. Only then can we understand how these insights are either accommodated or fail to be accommodated in risk analysis methods such as PRA.

1.1 The benefit of hindsight

As described by Kemeny (see [1], p. 43) and Rogovin (see [2], vol. 1, p. 12), the core melt accident at Three Mile Island started when an operator tried to clear a plugged system that supplied water to the primary reactor heat removal system used in electricity production (see [3], for decay heat process details). The operator became involved when a valve installed for the purpose of bypassing around the plugged system was not open ([1], pp. 47–48). Although the actions taken by the operator to clear the plugging were unsuccessful, the heat removal system was supplied with a backup protective system using a completely separate water supply. Unfortunately, the backup system's valves were inadvertently left shut, rendering it ineffective ([1], p. 47). The NRC requirements imposed through regulation and designed by the plant owners and investors anticipated such a loss of cooling sequence so a fourth separate protective system was supplied that would directly inject cooling water into the reactor core to keep it cool. Up to this point, the reactor system had begun heating up causing a relief valve to open, as designed, to reduce pressure.[2] Unfortunately, the relief valve failed to close when the pressure dropped back to normal, a malfunction that went unrecognized by the operators. However, the fourth method provided to keep the reactor core from overheating started automatically and began cooling the reactor core. Thus far in the accident sequence, the status of the reactor core protections could be summarized as: (a) main water supply—*plugged up*, (b) bypass protection of main supply—*valve shut*, (c) third water supply system—*valve shut*, (d) reactor system relief valve—*stuck open*, (e) fourth cooling system—*working and cooling*.

Because the relief valve stayed open too long, the reactor pressure kept dropping and the water around the core began expand as it boiled. Water from the reactor core was then pushed into a surge volume where the reactor system water level is measured. Although water was actually being continuously lost from the core through the stuck relief valve, the operators were under the impression, based on the surge volume level, that the reactor system was filling up by the fourth (and final) cooling system and they shut it off. At this point, the plant entered a state where it would lose the ability to prevent the reactor core from melting and releasing radioactive material. *An interesting detail is that even the malfunction of the relief valve sticking open was also anticipated and therefore a shutoff valve was provided; temperature measurements were put in place so the operators could understand if the relief valve was stuck open and they would know when the shutoff valve should be closed. The operators failed to close the valve even though the temperature measurement indicated they should* ([1], p. 46).

Although the several protections put in place would make it almost unimaginable the reactor core could melt, the NRC required a final protective system designed to protect the public in accidents releasing radioactive material from the reactor core. This final protection was a building that could withstand a very high pressure burst and remain

[2] Even this protective overpressure protective system was backed up with separate overpressure relief valves if the pressure had continued to rise.

effectively airtight. This building is called the "containment building" and it is the final barrier to release of radioactive material in the event of a serious accident.[3] Because the containment building was able to contain the radioactive materials until they could be properly managed, the public was not exposed to excessive radioactive material. The following considers what went wrong at Three Mile Island, what went right, and why?

1.1.1 What went wrong?

A complete description of the accident leading to core melt is given by Kemeny, but the main events could be summarized as: (a) the main cooling water supply was plugged up, (b) the backup system to keep the main water supply flowing was shut off, (c) the third backup system cooling water supply systems' valves were shut, (d) the reactor pressure relief system (itself a protection against overpressure) caused the reactor system to lose water due the valve sticking open and, (e) the fourth backup water system was shut off due to factors such as: incorrect data interpretation; unobserved data or; misunderstanding the thermodynamics of isenthalpic expansion.

Actions by the operator working on the main water supply system caused the main flow path to shut off (an "initiating event", see Sections 1.4, 3.2 and 5.3 and definition 5.2). The system that plugged up is added for chemical cleanup of the main water flow (see [1], p. 94). Although the possibility the chemical cleanup system could clog up was anticipated and protection had been added, the protective system failed to work because its bypass valve was not open; (see [1], p. 47–48), (see [2], vol. 1). The backup system with its valves shut (third main cooling system) did not operate for at least 8 min after it was needed. However this may not have directly made a difference in the core melt outcome (see [1], pp. 47, 91, 94). The reactor system relief valve opened at the right time but failed to close when the pressure had returned to normal; as a consequence, water continued to be released from the reactor system (see [1], pp. 28–29, 90–94). The fourth backup system that could have prevented the melt sequence started automatically and was working well but it was shut off by the operators (see [1], pp. 28, 89, 91, 93–94).

1.1.2 What went right?

As the accident progressed to melt, radioactivity monitors inside the plant began to show increasing levels of radiation. The radiation levels plus indications of increasing levels of water inside the reactor containment building gave the operators indications that an abnormal condition was present and worsening. Other indications, although working properly, were not seen or interpreted incorrectly by the operators thereby allowing the sequence to progress to core melt. By properly interpreting indications and taking appropriate actions in response, the operators may have kept the event from progressing. Even though instrument indications were "pointing" them to the possibility a dangerous event was in progress, the operators consistently failed to interpret critical indications correctly. It can be said that despite several backup systems and indications, the final most important element that went right was the NRC required a containment building acting as the final, in this case the fifth, barrier against radioactive material exposure to the public.

---

[3] The General Design Criteria require specific performance characteristics of containments. See, 36 FR 3256, 36 FR 3256, Feb. 20, 1971, as amended at 36 FR 12733, July 7, 1971; 41 FR 6258, Feb. 12, 1976; *43 FR 50163, Oct. 27, 1978*; 51 FR 12505, Apr. 11, 1986; *52 FR 41294, Oct. 27, 1987; 64 FR 72002, Dec. 23, 1999; 72 FR 49505, Aug. 28, 2007.*

1.1.3 Why?

The things that went wrong at Three Mile Island were mostly attributable to human errors of commission or errors of omission brought on by equipment breakdowns: (a) error of commission, the failed attempt to unplug the main water line, (b) error of omission, the failure to open the bypass valve, (c) error of omission, leaving the emergency water line valves shut, (d) error of commission, turning off the fourth protective system, (e) the reactor system relief stuck open and, (f) the main water supply plugged up. That is, the operators either took action that was wrong, error of commission, or did not take action when required, error of omission. These errors can be attributed to causes such as lack of training, lack of engineering insight, carelessness, or simply not following procedures. These kinds of errors are difficult to eliminate even when attempts are made to overcome them. For example, a second check of the required valve position would have been a good idea for the emergency valves; however this method is not always effective.[4]

1.2 Probability, reality, and maintenance policy

One may be tempted to look back at the Three Mile Island accident and argue the probability for the accident was extremely unlikely. Acting on such temptations that is, applying ex-ante probabilities to ex-post observations, is improper. The accident at Three Mile Island Unit 2 actually happened rendering the ex-post probability irrelevant. Prior to the accident some experts would say that the probability was very small, possibly on the order of 1 chance in a million. Of the many scenarios anticipated in experts' analyses of an ex-ante probability, the exact scenario that unfolded was unanticipated making their analyses irrelevant. Personnel misinterpreting indications, the actions taken due to the particular relief valve failure, and errors of commission were unanticipated *as a scenario*. The procedure for working on the chemical system could have required the operator to place the second protective system in service before starting the work as noted by Kemeny on page 47 of his report. In fact there are effectively limitless scenarios that may play out in the future. Solberg and Njå [4] state in their article "Reflections on the ontological status of risk", only one of the many (truly infinite) possible scenarios can play out to produce a current "state of affairs":

> *An important observation when we consider change or events is that when an event (a specified state of affairs) is manifested (when it happens) all other logically possible states of affairs are simultaneously excluded from manifestation. For this to be the case it would imply that there exists a whole range of possible future states of affairs, but only one of these would manifest at some point in time (the present).*
>
> *(Solberg and Njå, 2012)*

The discussion above is meant to point out why the accident at Three Mile Island happened even though the NRC and the plant design engineers endeavored to put in place protective systems intended to prevent progression to core melt. Before the accident occurred, a concerned citizen might question the design engineer in a conversation like the following:

Citizen, "What if the main cooling system gets plugged up?"
Engineer, "We thought of that and put in a bypass valve".

---

[4] See for example, https://www.nrc.gov/docs/ML1418/ML14188A495.pdf

Citizen, "What if the bypass valve doesn't work?"
Engineer, "We also thought of that possibility and added a separate system to pump water in. We also know that if this happens, the reactor system will heat up causing pressure to increase. But before you ask, we added relief valves for this possibility"

Citizen, "What if the relief valve doesn't work, will the system over pressurize?"
Engineer, "We thought of that too and added extra safety valves that would prevent overpressure"

Citizen, "But even so, there is no more cooling system left"
Engineer, "We put in yet another separate system to pump water in the reactor system just in case the other systems don't work"

At this point the citizen might reasonably stop asking questions, satisfied the NRC and the plant engineers had really thought of everything that could go wrong *in the future*.

It is clear that when a substantial hazard is present in a technological system, many questions need to be asked and answered, and many scenarios considered in detail. At Three Mile Island, the particular hazard was the need to remove decay heat from the reactor system over a relatively long period of time. Even though many questions may be asked and answered, once a technological system is started, we take a leap into the unknown. No one can say for sure what future scenarios may play out. Some may lead to great loss, injury, or even death. The maximum level of harm that could result from various failures in a technological system is almost certainly known by the engineers who design it. On the other hand, unless large datasets are available, the numerical *probability value* they would would occur cannot be known. Depending on the value of the technological system to the social welfare, those valuable enough to be allowed by law are most likely subject to regulation designed to protect citizens' safety and health.

1.3 Protective systems

Technological systems required by regulation are referred to as "protective systems". Such protective systems overlay technological systems operated for the purpose of maximizing profit. They differ from production systems in that they do not create products with a utilitarian purpose. Instead, protective systems function to reduce the probability that citizens would need to backstop losses from risks taken in owner and investor activities that exceed the investors' asset value. No one, not those maximizing profit, not the regulator, not the engineers, and not any independent scientific authority can know the probability or likelihood that a technological system in use will or will not cause harm with exactitude.

Protective systems add to the cost of goods and services. In so doing, they will generally balance the reduction of profit margins against the price citizens are willing to pay for the goods and services produced. In competitive markets, profit maximizers will look for opportunities to reduce costs imposed by protective systems. A subtlety created by the process of cost reduction is that profit maximizers will tend to look at all costs that go against their revenue including production costs in maintenance and operation. A reasonable question would be "does it matter to citizens if the maintenance and operational costs of the *production systems* are reduced if citizens require the regulator to make sure the *protective systems* are maintained up to standards?" The subtlety created by reducing costs in technological systems other than protective systems is that protective systems are required to protect against harms more often than if the production systems were operating smoothly. Regarding the Three Mile Island accident, Kemeny points out that,

> *Review of equipment history for the 6 months prior to the accident showed that a number of equipment items that figured in the accident had had a poor maintenance history without adequate corrective action. These included . . . the condensate polishers.*
> *(Kemeny, 1979)*

Failure of the condensate polishers at Three Mile Island triggered the sequence of events leading up to the core melting.[5] This "trigger" is commonly referred to as an "initiating event"; the first event in a series of events that may end with unpleasant consequences. he importance of these initiating events is that the more often they occur, the more often the protective systems will be called upon to operate and because their failure probability can not be known, citizens should be concerned if the maintenance and operational costs of the production systems are reduced to the point more triggers occur.

Whenever a protective system allows an initiating event to progress to harm, it is made obsolete by revision that would address the root cause of its failure. The NRC made substantial changes to protective system requirements in the nuclear reactors it regulates following the Three Mile Island accident. No one can know the probability a technological system that has the potential for harm will, or will not, cause harm over its useful lifetime in the absence of substantial data sets where harm has occurred. This well-known principle is classically described by Cardano and Wilks [5] and more formally by Bernoulli [6].[6] As technological systems are operated, protective systems that overlay them should be revised as necessary in light of new information to the point the technological systems are judged to be less harmful than the harms they are designed to overcome. Because protective systems add cost to the production of goods and services produced by technological systems, there is a balance, driven largely by market forces, to be struck among the cost of production, profit margin, and protection.

1.4 Predictive modeling for safety: critical protections

The concept of PRA is introduced in here at a very high level. In practice, PRA and other methods used to quantify probabilities and frequencies, are fundamentally derived from logical descriptions of a technological system's response to an "upset", or initiating event. **Figure 1** is an example of such a logical description that envisions a technological system that "operates", Z, on an initiating event input, I, to produce "outputs", O. Z creates at least 2^n outputs by assigning probabilities that split I at each of n devices. The splits are conditional such that if $P_{(\cdot)}$ is the probability for failure of the device, one branch of the split is assigned $P_{(\cdot)}$, and the other $\overline{P}_{(\cdot)}$.

Such probability models, with various levels of sophistication, are used to characterize the efficacy of safety-critical protections.[7] Understanding protective

[5] A description of the condensate polishing system is given in [1]. It contains resin beads that chemically condition the water flowing through them and act like a filter susceptible to plugging up.

[6] The "Law of Large Numbers" was formally proved by Bernoulli.

[7] While there is a vast engineering literature touting non-probabilistic strategies for addressing uncertainty (fuzzy logic, interval arithmetic, Dempster-Shafer, etc), only probability theory has survived the unrelenting scrutiny of philosophers of mathematics as a viable theory of uncertainty. We embrace probability theory as the preferred logical framework for predictive modeling in engineering practice.

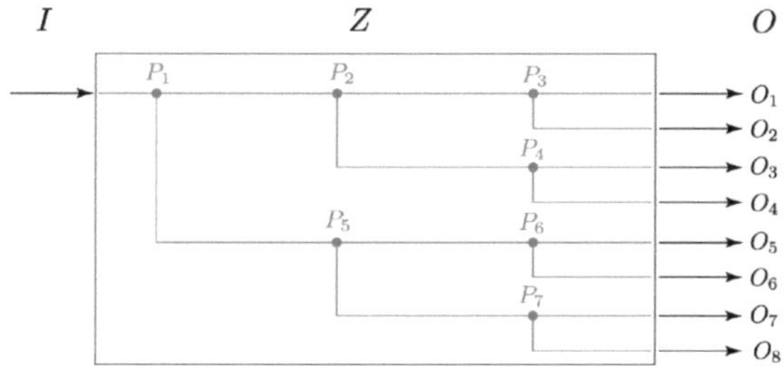

Figure 1.
A primitive event sequence logic for a protective system having three devices in parallel (1/3 logic).

system modeling discussed here necessarily assumes a postgraduate familiarity with stochastic processes distinguishing it from the treatment PRA practitioners use. Stochastic dynamics of safety-critical protections are defined in a general setting allowing treatment of efficacy analysis in a top-down manner starting at general predictive modeling followed by exploration of consequences inherent in the simplifying assumptions required of popular risk quantification methods such as PRA, Quantitative Risk Assessment (QRA), and Probabilistic Safety Assessment (PSA).

Models of efficacious protective system design and management cannot be informed by "probability numbers"—such numbers are unavailable as a practical matter. This is to say, the condition "safe enough" is almost never expressible as a numerical probability as argued by Hansson (see [7–9], for perspectives on safety). Hence, efficacy of protective systems should rightly be explored in terms of insights from the mathematical structure of stochastic processes serving as predictive models. To this end, the following develops models at a level of abstraction sufficient to capture stochastic structures that agree with practical engineering understandings. Most importantly, a top-down exposition leads to straightforward identification of the particularizing assumptions necessary to calibrate reactor protective system stochastic predictive models with observational data. In other words, the engineering physics that must hold to compute valid statistical estimates of risk metrics can be identified. Hansson in ([10], Section 4) sets the stage where statistical models can be calibrated and where they cannot.

Clearly, hazardous technologies and their protections are engineered systems that operate according to laws of physics, and their operational behaviors are governed by ongoing design, maintenance, and management decisions. These systems are carefully monitored over time so that information gained can direct re-designs and maintenance intended to improve productivity and safety. Of course, our uncertainty regarding how protections might perform in the future is, at best, a reflection of our historical understanding of physical behaviors, both technological and environmental. For example, we cannot forecast the future arrival of possible operational problems, the possibility of which we are *a priori* unaware.[8] Thus, any understanding of the efficacy of protections must connect physics with a time-dependent state of knowledge about that physics. In this regard, filtered probability spaces are indispensable when exploring engineered protective system efficacy.

[8] For example, one cannot assign the probability of occurrence of any failure mode that one does not yet know possibly exists ... a point seemingly overlooked by many risk analysts.

System physics are approached in here from an operations perspective to establish the analytical framework within which physical features can be mathematically described as time dependent functionals. This framework will be sufficiently general to accommodate most hazardous production systems. Analytical modeling of operations is concerned with characterizing a system's temporal dynamics as the evolution of a system's "state" over time. Almost always, *operations* is concerned with predicting, up to probability law, transitions among particular sets of system states given certain observable historical behaviors. Such transition dynamics are captured as stochastic processes.

The theory of stochastic processes is mature and provides a valid framework for understanding operations—derivations or detailed explanations of the central results applied in here are found in the widely accessible literature. Avoiding repetition, operation of hazardous systems and their protections are modeled in here by a stochastic point process and cited in the appropriate literature without proof.

2. Construction of accident counting processes for reactor risk analysis

Consider **Figure 2** showing how causality is aggregated from the underlying physics of nature, through the limitations of engineering understanding of the physics, up to the operational level where the physics is collapsed to observations *functional* or *failed* or, more succinctly, [0,1]. Risk quantification is done on observations whereby probabilities are assigned to events. Such probabilities are assumed in a probability space with events in sets of failure outcomes, $o_{(\cdot)}$, taking place at the level of the underlying physics shown in **Figure 2**.

2.1 Physics: based modeling in a general stochastic setting

Protective systems safeguard against future accidents. The design of protective systems necessarily relies on deterministic physical laws (physics) embedded within a stochastic framework that analytically captures uncertainty about *future* protection behaviors. This stochastic setting yields *predictive models* useful for understanding the efficacy of given protection designs. We establish the following:

Definition 2.1 (Object). A collection of devices, subsystems, systems, and environments enclosed within a specified logical control volume.

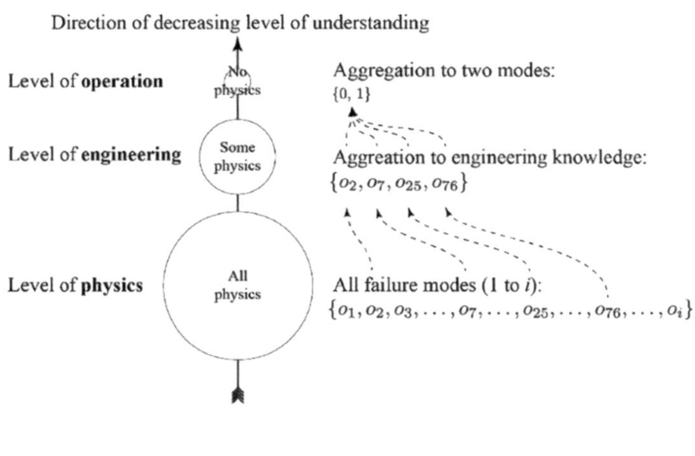

Figure 2.
Concept of knowledge aggregation from the underlying physics (both known and unknown) up to the operational level of observation.

Definition 2.2 (State). The unique numerical assignment to specific features of an object's physical condition (typically in SI units).

Definition 2.3 (State Space). The set of all possible object states.

Definition 2.4 (State Variable). A mapping from the domain of an object's physical condition into its state space.[9] Typically, state variables are indexed by time.

Definition 2.5 (State Trajectory). The evolution of an object's state over a collection of specific time intervals.

Definition 2.6 (Predictive Models). Probability laws with an object's state variables indexed by time. Predictive models typically include a time $t = 0$ that delineates future from past.

Remark 1 Predictive models *do not* predict the future. Rather, they frame uncertainty about future physical behaviors in terms of the present state of knowledge.

Objects of modeling interest can be comprised of subordinate objects; that is, an object can be the logical union of other objects. An object is *atomic* if it is not composed of subordinate objects. It follows that, an object's state space is defined on the union of its subordinate atomic object state spaces. The state of an object typically evolves over time. Hence, predictive modeling requires not only specifying the state variable mappings that assign state values to material and environmental conditions, but also crafting a mathematical characterization of the interaction of an object's state variables over time. The control volume defining any object connects its features with the laws of physics that govern its evolution, and it is our understanding of physics that allows us to mathematically characterize the time dependent interaction among an object's state variables. The time dependent behavior of an object's state is referred to as a *state trajectory*. The physical laws that govern state trajectories are, of course, specific to the object being modeled. Inasmuch as one's interests lie with establishing a mathematical framework within which predictive models for hazardous technologies can be crafted, they need only make reference to specific physics where application dictates.

Remark 2 Valid predictive models must respect two practical engineering constraints: 1) State trajectories are never physically observable beyond the present, and 2) Engineers are not clairvoyants. Valid predictive models are strictly informed by historical information, while allowing for the acquisition of additional information as the object's future unfolds (see [11], for an interesting perspective). The evolution of information history cannot be omitted from valid predictive models.

2.1.1 Life cycle, state trajectories, and the state process

The terminology *life cycle* is used when referring to object state trajectories that are defined over the entire open time interval $(-\infty, +\infty)$, giving a complete history of an object's state over time. In predictive modeling, an object's life cycle is rarely, if ever, identifiable with certainty at any finite time, and one must must be satisfied with expressing the likelihood that an observed state trajectory belongs to a specified collection of possible life cycles.

Clearly, an object's life cycle is determined by the physics within its control volume. In the usual manner, let the domain of an object's state variables be Ω, the collection of possible physical outcomes the object can experience. Further, let $\mathcal{F}$ be a σ-algebra on Ω such that $(\Omega, \mathcal{F})$ forms a measurable space.

[9] There is no guarantee that all object features in the domain of state variables are physically observable.

Elements of $\mathcal{F}$ are subsets of possible physical outcomes with a defined probability measure expressing the likelihood that a given subset of measurable outcomes contains an outcome of specific interest. P is a set function $P : \mathcal{F} \to \{0,1\}$, the triple $(\Omega, \mathcal{F}, P)$ forms a standard probability space (see [12], for a formal development).

Now, if Z_t is the object's state at time t, where $Z_t : (\Omega, \mathcal{F}) \to (S, \mathcal{B}(S))$, for $t \in \mathbb{R}$, then Z_t is a measurable state variable mapping.[10] If A is a measurable subset of $\mathbb{R}, Z_t(\omega) : \mathbb{R}_+ \to S$, and $t \in A$, then fixed $\omega \in \Omega$, forms a *state trajectory* under physical outcome ω. When $A = \mathbb{R}, Z_t(\omega)$ becomes the object's life cycle under outcome ω. Now, without loss of generality, require that for all $\omega \in \Omega$ and $t \in \mathbb{R}, Z_t(\omega)$ be right-continuous with left-hand limits. It then follows that the collection of random variables $\{Z_t; t \geq 0\}$ forms a stochastic process, with state space S. $\mathbf{Z} = \{Z_t; t \geq 0\}$ is referred to as an object's *state process*.

2.1.2 Probability law and validation

Predictive modeling requires framing uncertainty about an object's state through a *probability law* $\mathcal{L}_{\mathbf{Z}} : \mathcal{B}(S) \to [0,1]$ on its state process. Here, for each n-sequence two-tuple $\underline{B}_{\underline{t}_n} = ((t_1, B_1), \ldots, (t_n, B_n))$ such that $t_i \in \mathbb{R}_+$ and $B_i \in \mathcal{B}(S), i = 1, \ldots, n < \infty$,

$$\mathcal{L}_{\mathbf{Z}}\left(\underline{B}_{\underline{t}_n}\right) \equiv P\left(\bigcup_{i=1}^{n} Z_{t_i}^{-1}(B_i)\right) = P\{\omega \in \Omega : Z_{t_1}(\omega) \in B_1, \ldots, Z_{t_n}(\omega) \in B_n\}. \quad (1)$$

The measurability of all state variables in $\mathbf{Z}$ ensures that its probability law $\mathcal{L}_{\mathbf{Z}}$ defines a probability measure on the measurable space $(S, \mathcal{B}(S))$. Hence, uncertainty associated with any physics-based predictive model is captured on the probability space $(S, \mathcal{B}(S), \mathcal{L}_{\mathbf{Z}})$.

Definition 2.7 (Predictive Model). *For an object with state process* $\mathbf{Z}$ *defined on* $(\Omega, \mathcal{F}, P)$, *the probability space* $(S, \mathcal{B}(S), \mathcal{L}_{\mathbf{Z}})$ *is the object's predictive model.*

Validity of a specific predictive model $(S, \mathcal{B}(S), \mathcal{L}_{\mathbf{Z}})$ requires that it agree with both the physics underlying $\mathbf{Z}$ and historical observations of physical behaviors of the object that $\mathbf{Z}$ represents. Without crawling deeply into the weeds of model validation, it is easily reasoned that the degree of difficulty in proving validity is greatly influenced by the cardinality of both S and $\mathcal{B}(S)$. Suffice it to say that large state spaces associated with *complicated* object trajectories and *large* information bearing σ-algebras $\mathcal{B}(S)$ require vast amounts of observational data to prove model validity. In fact, validating $\mathcal{B}(S)$ is almost always a practical impossibility, since quantifying the probability law $\mathcal{L}_{\mathbf{Z}}$ stands among the most challenging aspects of stochastic modeling. However, careful stochastic analysis provides a direct means of identifying *invalid models*.

Observation 1 A necessary condition for an object's state process to be valid is that it must be congruent with well-understood physics. Conversely, in circumstances where $\mathbf{Z}$ is not congruent with well-understood physics, it immediately follows that the object's predictive model $(S, \mathcal{B}(S), \mathcal{L}_{\mathbf{Z}})$ must be invalid.

Observation 1 reveals the importance of stochastic modeling in a general setting where an object's *physics–based* behaviors are mapped into time–dependent state trajectories. While proving validity of a predictive model is rarely an achievable engineering objective, it is often straightforward to recognize invalidity of specific models.

---

[10] $\mathcal{B}(S)$ is the standard Borel σ-algebra generated by the object state space S.

The following Section 2.2 points out how popular risk modeling methodologies for protective system *operations* can easily obscure underlying physics and can result in a failure to recognize model invalidity.

2.2 Stochastic modeling: production, protection and environment

When concern lies with the efficacy of protections, **Z** is understood to model an object representing a hazardous production system protections with S as the object state space, onto which all underlying physics is mapped. Designate three subordinate objects as: a production system, a protective system, and an environment.[11] Without loss of generality, assume that these three subordinate objects are mutually exclusive (i.e., their respective control volumes do not intersect); mutual exclusivity does *not* imply that production, protection, and environment operate independently.

The mutual exclusivity of production, protection, and environment ensures that state space S is formed as the cross product of the subordinate object states. Here,

$$S = S^X \times S^R \times S^Y, \tag{2}$$

where
S^X is the subspace of production system states,
S^R is the subspace of protection system states, and.
S^Y is the subspace of environment states.

Take the subspaces identified in Eq. (2) to each be a manifold with boundary. By definition, any element of S^X numerically characterizes a production state. Similarly, elements of S^R quantify protection states, while subsets of S^Y describe an environmental state within which production and protection exist. Now,

$$\begin{aligned}
&\mathbf{X} = \{X_t : t \geq 0\}, \text{is the production state process, where}\\
&\quad X_t : (\Omega, \mathcal{F}) \to \left(S^X, \mathcal{B}\left(S^X\right)\right) \text{ is the production system state at time } t;\\
&\mathbf{R} = \{R_t : t \geq 0\}, \text{is the protection system state process, where}\\
&\quad R_t : (\Omega, \mathcal{F}) \to \left(S^R, \mathcal{B}\left(S^R\right)\right) \text{ is the protection system state at time } t; \text{ and}\\
&\mathbf{Y} = \{Y_t : t \geq 0\}, \text{is the environment state process, where}\\
&\quad Y_t : (\Omega, \mathcal{F}) \to \left(S^Y, \mathcal{B}\left(S^Y\right)\right) \text{ is the protection system state at time } t.
\end{aligned}$$

And, by construction $\mathbf{Z} = (\mathbf{X}, \mathbf{R}, \mathbf{Y})$, and for all $t \geq 0$

$$Z_t = (X_t, R_t, Y_t) : (\Omega, \mathcal{F}) \to (S, \mathcal{B}(S)) \tag{3}$$

where

$$\mathcal{B}(S) = \mathcal{B}\left(S^X\right) \otimes \mathcal{B}\left(S^R\right) \otimes \mathcal{B}\left(S^Y\right). \tag{4}$$

[11] Recall that a protective system is comprised only of elements that are mandated by regulatory oversight (i.e., its elements would not be present within the enterprise operations control volume in the absence of regulatory requirements).

2.2.1 Dependency among production, environment, and protection

By convention, take all trajectories of **X**, **R**, and **Y** to be càdlàg (*right-continuous* with *left-hand limits*). S partitioned with S^X, S^R, and S^Y allows framing the interaction among production, protection and environment over time. It is understood that the temporal dynamics of these three stochastic state processes are not mutually independent. Here, with the σ-algebras respectively generated by the processes **X**, **S**, **Y**, and **Z** written as

$$\mathcal{F}^X \equiv \sigma(\mathbf{X}) = \lim_{t\to\infty} \mathcal{F}_t^X = \sigma\{X_u; -\infty < u \le t\}, \tag{5}$$

$$\mathcal{F}^R \equiv \sigma(\mathbf{R}) = \lim_{t\to\infty} \mathcal{F}_t^R = \sigma\{R_u; -\infty < u \le t\}, \tag{6}$$

$$\mathcal{F}^Y \equiv \sigma(\mathbf{Y}) = \lim_{t\to\infty} \mathcal{F}_t^Y = \sigma\{Y_u; -\infty < u \le t\}, \tag{7}$$

$$\mathcal{F}^Z \equiv \sigma(\mathbf{Z}) = \lim_{t\to\infty} \mathcal{F}_t^Z = \sigma\{Z_u; -\infty < u \le t\}. \tag{8}$$

There can exist some $B_Z = B_X \cup B_R \cup B_Y \in \mathcal{F}^Z$, where $B_X \in \mathcal{F}^X, B_R \in \mathcal{F}^R$ and $B_Y \in \mathcal{F}^Y$ such that

$$P(B_X \cap B_R \cap B_Y) \neq P(B_X) \cdot P(B_R) \cdot P(B_Y). \tag{9}$$

Dependence among **X**, **R**, and **Y** is intuitively clear: The production state **X** and the protection state **R** both depend on **Y**, the common environment within which they operate. So, there is some $B \in S$ such that the likelihood that production and protection are in B at time $t \ge 0$ must as a matter of physics depend on the evolution of their common operating environment. For example the arrival of natural disasters and the wear-out rate of equipment are physics-based environmental influences on both production and protection. Thus,

$$\begin{aligned} P(X_t \in B | Y_u; u \le t) &\neq P(X_t \in B) \\ P(R_t \in B | Y_u; u \le t) &\neq P(R_t \in B), \end{aligned} \tag{10}$$

and, consequently, **X**, **R** are not generally independent. However, the future environment process **Y** is typically uninfluential on the histories of production and protection; we call this the Lack of Anticipation Property (LAP).

Definition 2.8 (Lack of Anticipation Property)**.** For all $t, s \ge 0$ and $B \in \mathcal{B}(S)$,

$$P(Y_{t+s} \in B | X_u, R_u : u \le t) = P(Y_{t+s} \in B) \tag{11}$$

The LAP plays a central role in predictive modeling of safety-critical protections, and is represented schematically in **Figure 3**.

2.2.2 Information flow and filtrations

Time and state dependencies among **X**, **R** and **Y** are informed by the flow of historical information. Information is characterized in a time indexed collection of sub-σ-algebras $\{\mathcal{F}_t\}_{t\ge 0}$, often called a history, forming a filtration augmenting the standard probability space on which state variables are defined. Here, the filtered

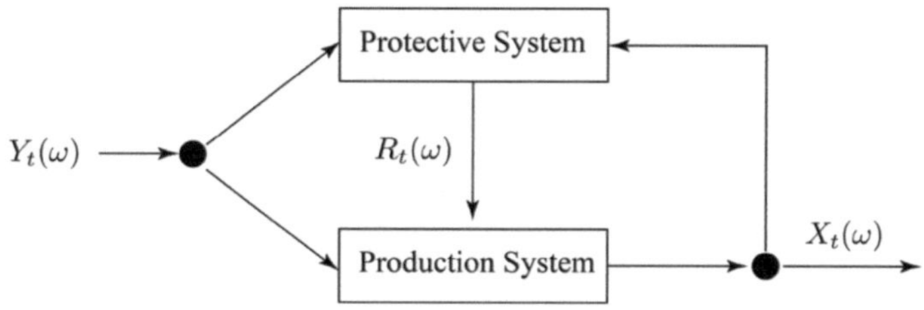

Figure 3.
Schematic of dependence among production, protection, and environment trajectories.

probability space $(\Omega, \mathcal{F}, \{\mathcal{F}_t\}_{t\geq 0}, P)$ is such that the state variables $Z_t = (X_t, R_t, Y_t)$ are $\mathcal{F}_t$-measurable for all $t \geq 0$. This filtration has the standard properties where,

Property 1 $\mathcal{F}_0$ contains all P–null sets (completeness).

Property 2 $\lim_{s\downarrow 0} \mathcal{F}_{t+s} \triangleq \cap_{s>t} \mathcal{F}_s = \mathcal{F}_t$ (right–continuity).

Property 3 $\mathcal{F}_t \subseteq \mathcal{F}_{t+s}$ (monotonicity).

Property 4 $\lim_{t\to\infty} \mathcal{F}_t \triangleq \vee_{t\geq 0} \mathcal{F}_t = \mathcal{F}$ (convergence).

The first property is an easily satisfied technical requirement from measure theory. The remaining three properties are normative and intuitively understandable. Property 2 indicates that the acquisition of engineering information is not necessarily *smooth*, with no information acquired in certain time intervals, while large amounts can be acquired at certain points in time. Property 3 asserts that information gained through discovery is not lost over time. Property 3 and Property 4 acknowledge that one cannot be convinced that all useful modeling information will be revealed prior to the end of lifecycles.

All information (engineering design, maintenance, operations, weather, management, *etc.*) associated with all possible object life cycles is contained in $\mathcal{F}$. Information available at time $t < \infty$ is limited to the sub-σ-algebra $\mathcal{F}_t$. Note that information represented in $\mathcal{F}_t$ is not necessarily limited to $\mathcal{F}_t^Z$, the σ-algebra generated by $\{Z_u; u \leq t\}$ (for example, maintenance schedules are not directly represented in $\mathcal{F}_t^Z$, but they are extremely useful in predictive modeling). Thus, in general, $\mathcal{F}_t^X$, $\mathcal{F}_t^R$, and $\mathcal{F}_t^Y$ are each sub-σ-algebras of $\mathcal{F}_t^Z$, and $\mathcal{F}_t^Z \subset \mathcal{F}_t$, for all $t \geq 0$.

Remark 3 Predictive modeling must rely on presently available information. With $t = 0$ taken as the beginning of all lifecycles in Ω, the sub-σ-algebra $\mathcal{F}_t$ contains all modeling information acquirable by time $t \geq 0$. The filtration $\{\mathcal{F}_t\}_{t\geq 0}$ contains all acquirable information flows over all possible complete lifecycles. $\{\mathcal{F}_t\}_{t\geq 0}$ is indispensable in predictive modeling because it enables modelers to represent the current state of knowledge. Omitting filtrations from predictive modeling implies that all possible modeling information is available at time $t = 0$; that is, one assumes that $\mathcal{F}_t = \mathcal{F}$ for all $t \geq 0$. The filtered probability space $(\Omega, \mathcal{F}, \{\mathcal{F}_t\}_{t\geq 0}, P)$ allows modelers to characterize their uncertainty about state variable values as a conditional (on available information) probability. For example with $t, s \geq 0, P(Z_{t+s} \in B)|\mathcal{F}_t)$ is the likelihood that the state of a reactor and its protection will be in condition $B \in \mathcal{B}(S)$ at time $t + s$, given the information available at time t.

2.2.3 Calibrating probability law

In recent years, Uncertainty Quantification (UQ) has emerged as a topic of considerable interest in risk analysis and safety engineering. Generally as the name suggests, UQ explores means to measure or judge the size of uncertainty. In the context of a nuclear reactor with regulated safeguards, operators and regulators are concerned with the efficacy of protection, and the predictive model $(S, \mathcal{B}(S), \mathscr{L}_Z)$ provides them a functional relationship between life cycle physics and uncertainty. Without exception, any useful UQ metric $m(\cdot)$ maps probability law into a real-valued n-vector; that is, $m : \mathscr{L}_Z \to \mathbb{R}^n, n \geq 1$. With $B \in S$ taken as the support of m, it is required that m be $\mathcal{B}(B)$-measurable. This mild measurability restriction admits UQ metrics that might not require a complete characterization of the probability law $\mathscr{L}_Z$. Nonetheless, all UQ metrics require that (if not all) at least some part of $\mathscr{L}_Z$ be calibrated. Calibrating $\mathscr{L}_Z$ is generally quite challenging.

Recall from the Kolmogorov Extension Theorem that specifying the probability law $\mathscr{L}_Z$ is equivalent to specifying all finite joint distributions on the state process $\mathbf{Z} = \{Z_t; t \geq 0\}$ [13]. Of course, there is generally an uncountable number of finite joint distributions on stochastic processes that evolve over continuous time. So, unless the state process $\mathbf{Z}$ exhibits very special independence properties (e.g., regeneration, stationarity, ergodicity) in combination with a small support for the UQ metric m, the amount of historical data required to statistically calibrate elements of $\mathscr{L}_Z$ is staggering.

Intuitively, crafting UQ metrics on predictive models that are useful for informing the efficacy of reactor protections boils down to:

1. Deciding how much detail a predictive model must include to support UQ metrics that modelers believe are useful in supporting reactor operation and public policy decisions.

2. Calibrating the needed elements of $\mathscr{L}_Z$ using historical information.

In practice, when particularizing $\mathbf{Z}$ to a specific reactor and safety–critical protective system, choosing a fine granularity on the support of any UQ metric $m : \mathscr{L}_Z \to \mathbb{R}^n$ introduces a model calibration burden for $\mathscr{L}_Z$ that is typically insurmountable. Section 4 reviews a collection of very strong underlying modeling assumptions required to calibrate Core Damage Frequency (CDF), the widely used UQ metric that is integral to the popular PRA methodology. The bottom line, here, is that UQ can never escape the challenges of calibrating $\mathscr{L}_z$; thus, UQ metrics applied to reactor protective system efficacy analysis should be evaluated with a healthy engineering mistrust.

3. Operations modeling

Operations generally refers to the study of *stochastic point processes* that are embedded within an object's state process $\mathbf{Z}$. It follows that in operations modeling and analysis *time* is the only physical variable of interest. An operations point process is typically embedded within $\mathbf{Z}$ and generates an increasing sequence of random times $\{T_n; n \in \mathbb{N}\}$ identifying the times of occurrence of a particular non-quantitative feature of $\mathbf{Z}$ that is identifiable through observable state changes. In principle, the state space S is covered with a partition $G \in \mathcal{B}(S)$, where G is an at most countable collection of sets. The elements of G correspond to important non-quantitative features on the state space S, and the random variable T_n marks the time of the nth occurrence of a state transition from one particular element of G to another.

In operations modeling there are practical measurability issues that must be considered when examining transitions among elements of the state space G by the trajectories of state process $\mathbf{Z}$ (which is, of course, adapted to the filtration $\{\mathcal{F}_t\}_{t\geq 0}$ of the filtered probability space $(\Omega, \mathcal{F}, \{\mathcal{F}_t\}_{t\geq 0}, P)$).

1. In principle, $Z_t^{-1}(G)$ must be $\mathcal{F}$-measurable for all $t \geq 0$.

2. There is no guarantee that $Z_t^{-1}(G) \in \mathcal{F}_t^Z$.

3. There is no guarantee that $Z_t^{-1}(G) \in \mathcal{F}_t$ (even when $Z_t^{-1}(G) \in \mathcal{F}$).

If $Z_t^{-1}(G) \notin \mathcal{F}_t$, then for all $t \geq 0$ there can exist important, and as yet undiscovered, operations events that may be observed in the future. The very practical circumstances where $Z_t^{-1}(G)$ is not $\mathcal{F}_0$-measurable are explored in Section 4.

In the remainder of this section, except where otherwise noted, the typical development of operations modeling where random quantities are defined on a standard probability space $(\Omega, \mathcal{F}, P)$, free from the information flow dynamics characteristic of filtered probability spaces, is followed.

Remark 4 Operations modeling is common in engineering practice. For example in reliability analysis $G = \{B, B^c\}$ partitions the state space S such that B is the collection of all *reliable* states. The standard indicator mapping $1_B : (S, \mathcal{B}(S)) \rightarrow (\{0,1\}, \sigma(\{0,1\}))$ of physical states into the set $\{0, 1\}$ defines a unit-less random variable on the probability space $(S, \mathcal{B}(S), \mathcal{L}_{\mathbf{Z}})$. The object's availability process $\mathbf{A} \triangleq \{A_t; t \geq 0\}$, where $A_t = 1_B(Z_t)$, has trajectories that proceed only in jumps (up or down) of magnitude one.[12] When T_n is taken as the time of the nth downward jump of $\mathbf{A}$, the sequence $\mathbf{T} \triangleq \{T_n; n \in \mathbb{N}\}$ corresponds to a stochastic point process marking time epochs of object state transition from *working* to *failed*. The failure-time process T is clearly subordinate to the state process $\mathbf{Z}$. *It is the point process* $\mathbf{T}$ *that characterizes object failures from an operations perspective.*

An operations point process T is often analyzed through its corresponding *counting* process $\mathbf{Q}$. Here,

$$\mathbf{Q} \triangleq \{Q_t; t \geq 0\}, \tag{12}$$

where,

$$Q_t = \sum_{n=1}^{\infty} 1_{[0,t]}(T_n) \tag{13}$$

counts the number of operations epochs appearing in the closed time interval $[0, t]$. There is an obvious one-to-one correspondence between the sequence $(T_n(\omega); n \in \mathbb{N})$ and the trajectory $Q_t(\omega), t \geq 0)$ for each $\omega \in \Omega$ – a plot of one directly reveals the other. It easily follows that, for all $n \in \mathbb{N}$ and $t \geq 0$, $\{Q_t \leq n\} = \{T_n \geq t\}$ which directly implies the distributional relationships

$$\begin{aligned} P(Q_t \leq n) &= 1 - P(T_n < t) \\ E[Q_t] &= \sum_{n=1}^{\infty} P(T_n \leq t). \end{aligned} \tag{14}$$

---

[12] The object's (limiting) availability often a metric of interest and is taken as $A \triangleq \lim_{t \to \infty} E\left[A_t | \mathcal{F}_t^Z\right]$, when it exists.

The operations point process **T** analysis is approached through its corresponding counting process **Q** which is accessible through the *martingale calculus*.

3.1 Classification of states for reactor operations with protections

Consider now an object representing a nuclear reactor, its regulated protections, and the environment within which the reactor and protections operate. The state process **Z** is defined, as before, in Section 2.2.2. Operations modeling and analysis begins with partitioning the state space S according $G = \left\{G_p, G_p^c\right\}$ where $G_p \in \mathcal{B}(S)$ is the set of all *persistent* states and its complement G_p^c is called the *transient* states. The set G_p is closed under all trajectories. That is, for all $s \in G_p$ and $\omega \in \Omega$

$$\lim_{t \to \infty} \sum_{0 < u \leq t} 1_{G_p}(Z_u(\omega)) \cdot 1_{G_p^c}(Z_{u-}(\omega)) = 0; \tag{15}$$

thus, once entering the set of all persistent states G_p, a trajectory $Z_t(\omega)$ cannot depart the set. $s \in G_p$ is an *absorbing* state if for all $\omega \in \Omega$

$$\lim_{t \to \infty} \sum_{0 < u \leq t} 1_{\{s\}}(Z_u(\omega)) \cdot 1_{\{s\}^c}(Z_{u-}(\omega)) = 0. \tag{16}$$

Clearly, absorbing states are persistent.

As a practical matter, require that all trajectories $Z_t(\omega)$, $\omega \in \Omega$ and $t \geq 0$, of the state process terminate with retirement.[13] Retirement occurs as either, (1) an inconsequential cessation of production, or (2) the consequence of an accident from which the reactor cannot recover. Once the reactor trajectory enters a retirement state, production terminates forever. All retirement states are taken as *absorbing*. Thus the absorbing states, designated $R \subset S$, and transient states designated R^c, partition S. The transients states R^c can themselves be partitioned such that N are *normal* operating states and D *distressed* operating states. Clearly, the sets $\{N, D, R\}$ also partition S. The possible state transitions among the elements of this partition are shown in **Figure 4**.

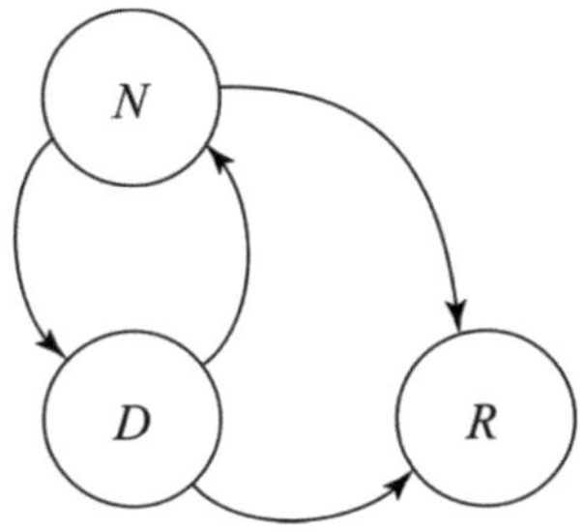

Figure 4.
Partitioned state transition diagram for state process **Z**.

[13] In sharp contrast to our analyses, PRA, QRA and PSA require that the state space S be almost surely (a.s.) composed of only persistent states none of which are not absorbing. The practical consequence of this state classification requirement is explored in Section 5.

3.2 The initiating event counting process

Recall that state variables are the three-tuple formed by $Z_t \triangleq (X_t, R_t, Y_t)$. Thus, when a trajectory $X_t(\omega), \omega \in \Omega$ and $t \geq 0$, of the production state process leaves the normal operating states $N \cap S^X$ and enters the distressed states $D \cap S^X$, an *initiating event* has occurred. Recalling that by convention all state trajectories are right-continuous and thus càdlàg, all initiating events occur at the exact instant of transition from $N \cap S^X$ to $D \cap S^X$. Arrival of an initiating event indicates that the reactor protective system should engage so as to mitigate potential harm that might arise with the reactor operating in the distressed states. In practice, initiating events are often (although not always) observable.

For all $\omega \in \Omega$ and $t \geq 0$, define

$$dQ_t(\omega) \triangleq 1_{\{D \cap S^X\}}(X_t(\omega)) \cdot 1_{\{N \cap S^X\}}(X_{t-}(\omega)). \tag{17}$$

Clearly for $t > 0, dQ_t : \Omega \to \{0, 1\}$ is a random variable on the measurable space of possible life cycles $(\Omega, \mathcal{F})$. For all $\omega \in \Omega$ and $t \geq 0$, it follows that

$$Q_t(\omega) = \int_{(0,t]} dQ_u(\omega) \triangleq \sum_{0 < u \leq t} 1_{\{D \cap S^X\}}(X_u(\omega)) \cdot 1_{\{N \cap S^X\}}(X_{u-}(\omega)) \tag{18}$$

form the trajectories of the stochastic process $\mathbf{Q} = \{Q_t; t \geq 0\}$ inherit the càdlàg property. We call $\mathbf{Q}$ the *initiating event counting process*. Here, as a practical consideration, we will require that $Q_0(\omega) = 0$ for all $\omega \in \Omega$. That is, we assume that reactor production should be in a normal operating state at time $t = 0$ at the beginning of lifecycle ω.[14]

It is a straightforward matter to construct the operations point process $\mathbf{T} = \{T_n; n \in \mathbb{N}\}$ which captures the arrival times of initiating events. $\mathbf{T}$ is referred to as the *initiating event process*, and for each $\omega \in \Omega$ and $n \in \mathbb{N}$,

$$T_n(\omega) \equiv \inf_{t > 0} \left\{ \int_{[0,t]} dQ_s(\omega) = n \right\} \leq \infty. \tag{19}$$

The initiating event counting process $\mathbf{Q}$ plays a central role in our treatment of operations modeling and efficacy analysis of safety-critical reactor protections. Understanding the construction of $\mathbf{Q}$ provides important insights (to be discussed in subsequent sections) into the relationship between hazard, risk, and the efficacy.

Remark 5 When assuming that $Z_t^{-1}(G) \in \mathcal{F}_0$ and with $Q_t : (\Omega, \mathcal{F}^Z) \to (\mathbb{N}, \mathcal{B}(\mathbb{N}))$, by construction it is ensured that Q_t is $\mathcal{F}_t$–measurable for all $t \geq 0$; hence, the initiating event process $\mathbf{Q}$ is adapted to the filtration $\{\mathcal{F}_t\}_{t \geq 0}$ which contains the history generated by the state process $\mathbf{Z}$. It is preferable to model operations processes on the

[14] Support for this assumption requires the state (normal) to be observable.

filtered probability space $(\Omega, \mathcal{F}, \{\mathcal{F}_t\}_{t\geq 0}, P)$, because its filtration contains all available information ... not simply the natural filtration $\{\mathcal{F}_t^Q\}_{t\geq 0}$ of **Q**. When PRA, QRA and PSA are used for risk analysis, they implicitly and strictly rely on the natural filtration

$$\{\mathcal{F}_t^Q\}_{t\geq 0} \triangleq \{\sigma(Q_t)\}_{t\geq 0} \subset \{\mathcal{F}_t\}_{t\geq 0} \tag{20}$$

which contains only information about the occurrence times of initiating events, while ignoring all history of the state process **Z**, maintenance activity, weather, *etc*.

3.3 The accident counting process

When safety-critical protections function properly a distressed reactor (i.e., $X_t \in \{D \cap S^X\}$) will avoid an accident and return to the set of normal operating states in N. Accidents are events influencing states of nature outside the control volume within which a nuclear reactor's state process **Z** develops. Accidents are always a consequence of protection failure that causes collateral harm.[15] For our purposes, when an initiating event leads to any level of collateral harm, then that event becomes an epoch of an accident.[16] Possible time delays are allowed for between the arrival of an initiating event and any eventual collateral harm.[17] To this end, distressed state $D \subset S$ is partitioned into those states C that impact physics outside the system control volume causing collateral harm and $E = D/C$ those distressed states that do not cause collateral harm. Partitioning D allows an operations characterization of an accident:

Definition 3.1. An **Epoch of accident** occurs when the trajectory $(X_t, R_t)(\omega), \omega \in \Omega$ is such that the production state process **X** enters $\{C \cap S^X\} \subset D \subset S$ while the state of the protection process **R** is in $\{D \cap S^R\}$.

Guided by the transition diagram of **Figure 5**, a modification of **Figure 4**, showing the partitioning of distressed states D into C and E, the accident counting process and the accident point process can be constructed.

Remark 6 Note that in the interval following an initiating event indicating that the state process has moved into distress, the system state can possibly transition many times between accident states in C and non-accident distressed states in E before either returning to normal or being retired. In practical modeling scenarios, state transitions across the partition of distressed states might take months (or even years). Thus, the initiating event leading to distress might not exceed to an accident for quite some time. In such situations, knowing that an initiating event has just occurred does not necessarily reveal whether or not the system state is on a trajectory of accident.

Now, define

---

[15] A good system for setting level of consequence is the International Nuclear Event Scale (INES). In this system, we would call events similar to those having INES Level 4 or above as an accident. The NRC defines an Extraordinary Nuclear Occurrence (ENO) in 10 CFR 140.83.

[16] Consideration of disastrous events is from a public policy perspective and focuses on collateral economic harm to people and/or the environment outside of the security fence.

[17] Ground water contamination sourced from the Savannah River Site facility continued developing over many years, with an accident being discovered long after the facility first enter into distressed states of operation.

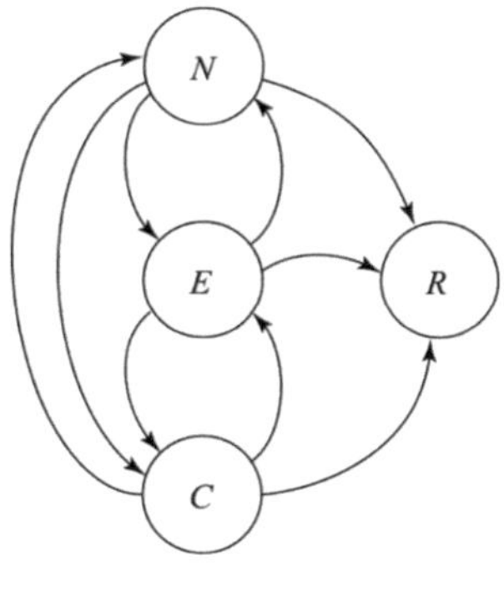

Figure 5.
Partitioned state transition diagram for the state process **Z** *with the distresses states* D *decomposed into* C *(catastrophic states) and* E *non-catastrophic distressed states.*

$$\begin{aligned} dQ_t^C(\omega) &\triangleq Q_t^C(\omega) - Q_{t-}^C(\omega) \\ &= 1_{\{C\cap S^X\}}(X_t(\omega)) \cdot 1_{\{D\cap S^R\}}(R_t(\omega)) \cdot 1_{\{N\cap S^X\}}(X_{t-}(\omega)). \end{aligned} \tag{21}$$

Note that the time $T_m(\omega)$ of the mth arriving epoch of an accident is given by

$$T_m^C(\omega) \triangleq \inf_{t>0} \left\{ \int_{(0,t]} dQ_u^C(\omega) = m \right\} \leq \infty. \tag{22}$$

When m is the index of the last arriving epoch of an accident before retirement, all subsequent accidents are taken by convention to occur at infinity. Clearly, the random sequence $\{T_m^C; m \in \mathbb{N}\}$ is a thinning of the initiating event process $\mathbf{T}$. The point process $\mathbf{T^C} \triangleq \{T_m^C; m \in \mathbb{N}\}$ is the *accident process*.[18]

With Q_t^C being the number of arriving accidents in the interval $(0, t]$, it follows that for each $\omega \in \Omega, t \geq 0,$ and $m \in \mathbb{N}$

$$Q_t^C(\omega) = Q_{T_m^C(\omega)}(\omega), t \in \left[T_m^C(\omega), T_{m+1}^C(\omega)\right), \tag{23}$$

and, in practice with $Q_t^C(\omega) = 0$ for all $\omega \in \Omega$,

$$\begin{aligned} Q_t^C(\omega) &= \int_{[0,t]} dQ_u^C(\omega) \\ &\triangleq \sum_{0<u\leq t} 1_{\{C\cap S^X\}}(X_u(\omega)) \cdot 1_{\{D\cap S^R\}}(R_u(\omega)) \cdot 1_{\{N\cap S^X\}}(X_{u-}(\omega)). \end{aligned} \tag{24}$$

$\mathbf{Q^C} \triangleq \{Q_t^C; t \geq 0\}$ is called the *accident counting process*. By construction, $\mathbf{Q^C}$ is adapted to the filtration of $(\Omega, \mathcal{F}, \{\mathcal{F}_t\}_{t\geq 0}, P)$.

Remark 7 When assuming that $Z_t^{-1}(G) \in \mathcal{F}_0$, the accident counting process $\mathbf{Q^C}$ is, clearly, subordinate the initiating event counting process $\mathbf{Q}$. That is, each epoch of arrival

[18] The arrival processes $\mathbf{T}$ and $\mathbf{T^C}$ play a prominent role in PRA, QRA, and PSA. However, studying their corresponding arrival counting processes in order to access important results from the martingale calculus is preferred.

in $\mathbf{Q}^{\mathrm{C}}$ is also an epoch of arrival in $\mathbf{Q}$. It is important to note, however, that while both $\mathbf{Q}^{\mathrm{C}}$ and $\mathbf{Q}$ are adapted to the natural filtration $\{\mathcal{F}_t\}_{t\geq 0}$, $\mathbf{Q}^{\mathrm{C}}$ *is not* adapted to the natural filtration $\{\mathcal{F}_t^Q\}_{t\geq 0}$ of the initiating event process $\mathbf{Q}$. Thus, the history of initiating events alone contains insufficient information to construct the accident thinning—a physics-based insight often overlooked in popular quantitative risk methodologies.

Finally, as a notational convenience, define for all $\omega \in \Omega$ and $t \geq 0$

$$A_t(\omega) \triangleq 1_{\{N\cap S^R\}}(R_t(\omega)), \tag{25}$$

where $A_t : (S, \mathcal{B}(S)) \rightarrow (\{0,1\}, \mathcal{B}(\{0,1\})$ is a random variable of the filtered probability space $(\Omega, \mathcal{F}, \{\mathcal{F}_t\}_{t\geq 0}, P)$. Clearly,

$$A_t = \begin{cases} 1, & \text{protections are available at time } t \\ 0, & \text{otherwise.} \end{cases} \tag{26}$$

We call $\mathbf{A} = \{A_t; t \geq 0\}$ the *protection availability process*, which inherits càdlàg properties from $\mathbf{R}$. Substituting Eq. (17) and Eq. (25) into Eq. (24) gives

$$\boxed{Q_t^C(\omega) = \int_{[0,t]} (1 - A_u(\omega))dQ_u(\omega)} \tag{27}$$

for all $\omega \in \Omega$ and $t \geq 0$. And, it follows directly from Eq. (22) and Eq. (27) that for all $m \in \mathbb{N}$, $\omega \in \Omega$ and $t \geq 0$

$$\boxed{T_m^C(\omega) \triangleq \inf_{t>0} \left\{ \int_{(0,t]} (1 - A_u(\omega))dQ_u(\omega) = m \right\} \leq \infty.} \tag{28}$$

The importance of Eq. (27) is that it gives the dynamic relationship between the number of accidents over time in terms of arriving initiating events and the reliability of protections. It should be clear from the construction of Eq. (27) from trajectories of the state process $\mathbf{Z}$ that the stream of arriving initiating events and the reliability of protections are *not* generally independent since both random phenomena are stochastically dependent on the dynamics of the environment process $\mathbf{Y}$. Further, Eq. (22) establishes the relationship between the accident time point process $\mathbf{T}^{\mathrm{C}}$ and initiation event arrivals and protection reliability where, again, stochastic dependence between initiating event arrivals and protection reliability cannot be ignored. Eqs. (22) and (27) play a central role in the developments presented in Sections 4 and 5.

4. Unknown-unknowns

When designing nuclear reactor protections, it is impossible to foresee and design for every circumstance that might lead to an accident. Such design deficiencies are often called *unknown-unknowns*.[19] Unknown-unknown design deficiencies are

---

[19] The terminology *unknown-unknowns* was popularized by former United States Secretary of Defense Donald Rumsfeld.

routinely discovered, documented, and corrected. It can be shown that uncertainty about the consequences of as yet undiscovered design deficiencies cannot be quantified. Thus, Probability Quantification (PQ) methodologies without exception overlook the influence of unknown-unknowns on initiating events, protection reliability, and ultimately epochs of accidents. Overlooking unknown-unknowns can only bias predictive accident metrics such as CDF optimistically, presenting an unavoidable pitfall for quantitative methodologies including PRA, QRA and PSA that make no use of filtered probability spaces.

Unknown-unknown failure modes in protective systems will be the focus of analytical developments in the following.[20] Clearly, the possibility of undiscovered failure modes in reactor protections is a matter of great concern to both operators and regulators. The NRC has established rigorous reporting and operation protocols focused on the discovery of newly discovered protection design flaws.

The analytical consequences of unknown-unknowns are revealed only when the system state process **Z** is defined on the filtered probability space $(\Omega, \mathcal{F}, \{\mathcal{F}_t\}_{t\geq 0}, P)$, where all predictive assertions must be addressed in the context of information currently available in the filtration $\{\mathcal{F}_t\}_{t\geq 0}$. In particular suppose that in the design of protections there exists protection failure modes that are undiscovered at the time of deployment. These design inadequacies will only be discovered during operations and are then incorporated into the existing body of information. Thus, protection failure modes unknown at time $t = 0$ will enter into the filtration $\{\mathcal{F}_t\}_{t\geq 0}$ at some time $t > 0$ upon discovery.

By construction as shown in Sections 2 and 3 the protection availability process **A** is adapted to the filtration $\{\mathcal{F}_t\}_{t\geq 0}$ (i.e., A_t is $\mathcal{F}_t$-measurable). When for example examining the likelihood that protections are not in a normal condition at any time $t \geq 0$, it follows that

$$\begin{aligned} P(A_t = 0) &= 1 - E[A_t] \\ &= 1 - E[E[A_t | \mathcal{F}_t]], \end{aligned} \tag{29}$$

because the random variable $E[A_t | \mathcal{F}_t]$ is well-defined.[21]

In predictive modeling, it is often the case that the condition of protections at some random time A_τ, where $\tau : \Omega \to \mathbb{R}^+$, is of great interest. Here, care must exercised because Eq. (29) does not necessarily hold when substituting A_τ for A_t. Let $G_f \subset S$ be the set of all states where protections are unavailable (protections can be either failed or out of service due to maintenance). Now for t taken at the present time, let τ be the time of the next arriving initiating event in the initiating event process **T**. It follows that

$$\tau = T_{Q_t+1} > t, \tag{30}$$

with Q_t being the number of initiating events arriving in the interval $[0, t]$. Now consider two cases: (1) $A_t^{-1}(G_f)$ is $\mathcal{F}_0$-measurable, and (2) $A_t^{-1}(G_f)$ is *not* $\mathcal{F}_0$-measurable.

---

[20] Unknown-knowns also bias predictive models of initiating events.

[21] $E[A_t | \mathcal{F}_t]$ represents the equivalence class of random variables satisfying the definition of conditional expectation on the probability space $(\Omega, \mathcal{F}, P)$.

4.1 Case 1: $A_t^{-1}(G_f) \in \mathcal{F}_0$

By definition, $G_f \in \mathcal{F}$ is the collection of all states in S where protection is unavailable. Thus, any failure mode for protection must be reflected as a state $s \in G_p \subset S$. When $G_p \in \mathcal{F}_0$, it follows that every possible failure mode must be known at time $t = 0$ since the pre-image of $G - f$ through the random variable A_t appears in $\mathcal{F}_0$. By definition the filtration $\{\mathcal{F}_t\}_{t \geq 0}$, $\mathcal{F}_0 \subset \mathcal{F}_t$ for all $t > 0$ implies that $A_t^{-1}(G_p) \in \mathcal{F}_t$ for all $t \geq 0$. It now follows that for any $u > 0$ $A_{t+u}^{-1}(G_p) \in \mathcal{F}_t$ and, since τ is an $\mathcal{F}_t$ stopping time, $\tau > t, A_\tau^{-1}(G_p) \in \mathcal{F}_t$. In other words, when all possible states where protection is unavailable (including protection failure modes) are understood at time $t = 0$, then

$$\begin{aligned} P(A_\tau = 0) &= 1 - E[A_\tau] \\ &= 1 - E[E[A_\tau | \mathcal{F}_t]] \end{aligned} \tag{31}$$

gives the likelihood that the next arriving initiating event will result in an accident.

4.2 Case 2: $A_t^{-1}(G_f) \notin \mathcal{F}_0$

Suppose now that there exist protection failure modes that are unknown at time $t = 0$. This implies that $A_t^{-1}(G_f) \notin \mathcal{F}_0$. When $A_t^{-1}(G_f)$ is not $\mathcal{F}_0$-measurable there can be no guarantee that $A_t^{-1}(G_f)$ will be $\mathcal{F}_t$-measurable for any $0 < t < \infty$. That is, for any $u > 0$, we cannot ensure that A_{t+u} is $\mathcal{F}_t$-measurable. And, because $\tau > 0$, we *cannot ensure* that A_τ is $\mathcal{F}_t$-measurable even though τ is an $\mathcal{F}_t$ stopping time. In fact there will always be some $t > 0$ where $A_\tau \notin \mathcal{F}_t$ which implies that $E[A_\tau | \mathcal{F}_t]$ is not well-defined. From the failure of $E[A_\tau | \mathcal{F}_t]$ to satisfy the definition of a random variable on Ω, it can be concluded that when there are undiscovered protection failure modes, Eq. (31) cannot hold.

4.3 The practical implications of undiscovered protection failure modes

It is reasonable at this juncture to consider the extent to which protection failure modes not identified in $\mathcal{F}_0$ are problematic. Let τ_d be the time of first discovery of a heretofore undiscovered protection failure mode. Some observations that can be normatively understood are:

- The discovery time τ_d of a specific heretofore undiscovered failure mode is random and measurable with respect to probability space $(\Omega, \mathcal{F}, P)$.
- τ_d is not an $\{\mathcal{F}_t\}_{t \geq 0}$ stopping time. Clearly, $\{\tau \leq t\}^c \notin \mathcal{F}_t$.
- Any protection failure mode will almost surely be found over the lifecycle of protections.
- It is possible that a failure mode is first discovered through accident postmortem analysis (i.e., the failure mode caused an accident upon its first appearance).
- For any time $t \geq 0$ a non-clairvoyant cannot rule out the possibility of remaining undiscovered protection failure modes.

- If undiscovered failure modes are in play, the filtration $\{\mathcal{F}_t\}_{t\geq 0}$ has not yet converged and all predictive PQ metrics must be formed conditioned on the current information represented in filtration $\{\mathcal{F}_t\}_{t\geq 0}$ ... which does not include all information represented in $\mathcal{F}$.

- Nearly all historical nuclear reactor accidents have been attributed to heretofore unanticipated protection failures.

If there is sufficient confidence in a non-clairvoyant belief that that all protection failure modes have been discovered, then all information impacting protection design will be found in the tail σ-algebra $\mathcal{T}$, where

$$\mathcal{T} \triangleq \bigcap_{t\geq 0} \sigma(\{A_u; u \geq t\}). \tag{32}$$

By definition, $\mathcal{T} \subset \mathcal{F}$ and characterizes design information in the *remote future* of **A**. In the remote future of protections, there can be no undiscovered failure modes.

Typically, PQ methodologies (e.g., PRA, QRA, PSA) implicitly rely on the assumption that events associated with protections are $\mathcal{T}$-measurable, because this assumption guarantees the existence of

$$A = \lim_{t\to\infty} A_t \quad \text{and} \quad \overline{A} = \lim_{t\to\infty} \frac{1}{t}\int_0^t A_u(\omega)du, \omega \in \Omega. \tag{33}$$

The expected value $E[A]$ and its statistical estimator $\overline{A}$ play essential roles in PRA, QRA and PSA, and they are only practically accessible under the assumption that all available information is in the tail σ-algebra $\mathcal{T}$. The estimator $\overline{A}$, when computed with data collected other than in $\mathcal{T}$ will always underestimate the true value of $E[A]$, a dangerous bias.

5. The hierarchy of modeling assumptions supporting PRA

PRA appears, today, in regulatory language, NRC directives and even in federal legislation. PRA methodology relies on very strong modeling assumptions that are rarely (if ever) explicitly qualified in practice. Owing to the wide acceptance and application of PRA in the civilian nuclear industry, it is useful to identify and explain the assumptions underlying this popular risk analysis methodology. These modeling assumptions are best explained hierarchically so as to reveal a sequence of increasingly strong and necessary conditions leading to the computation of CDF ... the central risk metric derived from PRA. These assumptions are difficult to justify in practice and failure to satisfy any of them leads to optimistically biased estimates for CDF.

Begin again with the state process **Z** characterizing the time dependent behavior of a reactor's production, protections, and common surrounding environment defined on the filtered probability space $(\Omega, \mathcal{F}, \{\mathcal{F}_t\}_{t\geq 0}, P)$, where the state variables are mapped onto the measurable state space $(S, \mathcal{B}(S))$. The state space S is partitioned according to **Figure 5** in Section 3.

Now identify seven modeling assumptions that must be enforced in order to calibrate CDF using historical data.

Assumption 1 There are no unknown-unknowns.

Assumption 2 There are no absorbing states.

Assumption 3 CDF is well-defined.

Assumption 4 Arriving initiating events see the time average of protection availability.

Assumption 5 Arriving initiating events form a Poisson process.

Assumption 6 The protection availability process is stationary and ergodic.

Assumption 7 Protection unavailability is independent of all initiating events.

Assumptions 1–7 are numbered such that each implicitly enforces all lower numbered assumptions. Assumptions 6 and 7 are required in order to calibrate CDF with historical data. The consequence of these cumulative assumptions are examined one-at-a-time and in order.

5.1 Cumulative assumption: there are no unknown-unknowns

As discussed in Section 4, assuming that there are no as yet undiscovered states requires that

$$Z_t^{-1}(B) \in \mathcal{F}_0, \tag{34}$$

for $B \in \mathcal{B}(S)$ and $t \geq 0$. Recall that by definition of the filtration $\{\mathcal{F}_t\}_{t\geq 0}$ $\mathcal{F}_t$-measurability is also ensured. This modeling assumption cannot be verified so long as a reactor is not retired. Obviously, with the random time $\tau : \Omega \to (0, \infty)$ taken as the time of discovery of the last heretofore undiscovered state, τ is not $\mathcal{F}_t$-measurable for any $t < \infty$, which implies that there cannot be an observable condition confirming that there are not more undiscovered states. Clearly, the no unknown-unknowns assumption reflects a (believed) near clairvoyant understanding of the technologies being modeled.

5.2 Cumulative assumption: there are no absorbing states

The absence of absorbing states implies that either all $s \in S$ are transient, or there exists at least one non-singular collection of persistent states $B_\alpha \in S, \alpha \in \mathcal{A}$ such that

$$P(\lim{}_{t\to\infty} Z_t \in B_\alpha) > 0, \tag{35}$$

and when a state trajectory enters B_α it never exits, visiting each state $s \in B_\alpha$, infinitely often. Further,

$$P\left(\lim{}_{t\to\infty} Z_t \in \bigcup_{\alpha\in\mathcal{A}} B_\alpha\right) = 1, \tag{36}$$

which implies that each state trajectory must eventually enter a non-singular collection of persistent states that it visits infinitely often.

The PRA methodology implicitly rejects the possibility that all states in S are transient, and therefore accepts that all state trajectories visit some collection of states infinitely often. Further, PRA some of the states visited infinitely often are accidents (else there would be no reason to perform PRA). It follows that accepting the assumption that reactors are never retired (either through an accident or cessation of operations), there will be accidents *ad infinitum*.

5.3 Cumulative assumption: CDF is well defined

Accepting Assumptions 1 and 2, it is feasible to investigate CDF. Here, as before, Q_t^C, counts the number of core damage events a reactor suffers in the interval $[0, t]$. CDF is a frequency and thus is the limiting number of core damage events per unit time. That is,

Definition 5.1 (Core Damage Frequency).

$$\text{CDF} = \lim_{t\to\infty} \frac{1}{t} Q_t^C \tag{37}$$

whenever convergence to a constant occurs a.s.

CDF is treated as a numerical constant. However, convergence of Eq. (37) is not guaranteed. Further, even when CDF converges, there is no guarantee that its limit is a constant; convergence to a random variable is a completely plausible circumstance. It must be emphasized that CDF is a numerical constant, estimates of which are used to gauge the risk of suffering a core damage event.

The existence of CDF is determined in part on the dynamics of initiating event arrivals. The limiting arrival rate λ of initiating events is defined as follows.

Definition 5.2 (Initiating Event Frequency).

$$\lambda = \lim_{t\to\infty} \frac{1}{t} Q_t \tag{38}$$

convergence occurs a.s. In general λ can be a random variable, and convergence to a constant occurs only when $\lambda = E[\lambda]$.

Recall that $\mathbf{Q}^\mathbf{C}$ and $\mathbf{Q}$ are right-continuous. It follows directly from definitions 5.1 and 5.2 that when λ is a constant,

$$\text{CDF} \overset{\text{a.s.}}{=} \lim_{t\to\infty} \frac{Q_t^C}{Q_t} \frac{Q_t}{t} = p\lambda \tag{39}$$

where, often the *split fraction* is

$$p \triangleq \lim_{t\to\infty} \frac{Q_t^C}{Q_t}. \tag{40}$$

Thus, CDF is the product of the initiating event frequency with the proportion of initiating events that exceed to core damage. Typically, estimating λ is straightforward. Estimating p is more challenging. Note that while p takes values in the interval $[0, 1]$, it is defined as the limiting value of a ratio of random variables. However, under certain circumstances p can be interpreted as the probability that an arriving initiating event will exceed to core damage. One such circumstance occurs when initiating events form an ordinary Poisson process. Then, the well-known Poisson Arrivals See Time Averages (PASTA) result applies, and p can be computed as the limiting unavailability of reactor protections (see [14]). Unfortunately, the assumptions needed to justify PASTA defy practical justification. Consequently, estimating p is quite difficult in practice.

There are a variety of approaches for crafting estimators of p that incorporate the joint histories of initiating event arrivals, protection maintenance activity, environmental conditions, etc. Monte Carlo methods, owing to their adaptability to complex engineering models, have gained acceptance and popularity for estimating p and other statistics. These methods are not, however, a panacea because they require characterizing probability laws on subordinate stochastic processes that must be mapped into the dynamics of protection availability in order to build useful estimators. Characterizing probability laws on stochastic processes is often impractical due to the intensive data support required for all but the most stylized processes.

In order to better appreciate the manner in which CDF jointly depends on the arrival of initiating events and the efficacy of reactor protections, we will appeal to the martingale characterization of stochastic point processes (see [15]). For practical considerations, we require that initiating events occur one at a time a.s.

Since the trajectories of $\mathbf{Q}$ are nonnegative non-decreasing and proceed in jumps of size one a.s., clearly $E[Q_{t+u}|\mathcal{F}_t] \geq Q_t$ for all $t, u \geq 0$. Thus, $\mathbf{Q}$ forms a sub-martingale on the filtered probability space $(\Omega, \mathcal{F}, \{\mathcal{F}_t\}_{t\geq 0}, P)$. Appealing to the standard results from the martingale calculus (see [16]), it now follows from the Doob-Mayer Decomposition Theorem that

$$Q_t = M_t + \Lambda_t, \tag{41}$$

where the process $\mathbf{M} \triangleq \{M_t; t \geq 0\}$ forms an $\mathcal{F}_t$–martingale with compensator $\{\Lambda_t; t \geq 0\}$. Hence,

$$E[M_{t+u}|\mathcal{F}_t] = M_t = 0 \tag{42}$$

for all $t, u \geq 0$, and Λ_t is increasing a.s. and $\mathcal{F}_t$–predictable, with

$$\Lambda_t = \int_{[0,t]} \lambda_u du. \tag{43}$$

Here, λ_t is well-defined whenever for all nonnegative, $\{\mathcal{F}_t\}$-predictable $\{C_t\}_{t\geq 0}$

$$E\left[\int_{[0,\infty]} C_t dQ_t\right] = E\left[\int_{[0,\infty]} C_t \lambda_t dt\right] \tag{44}$$

is satisfied. $\mathbf{M}$ is called the *initiating event martingale*.

When well defined, λ_t is a unique Radon-Nikodym derivative defined on the usual equivalence class, with the stochastic intensity process $\{\lambda_t; t \geq 0\}$ adapted to $\{\mathcal{F}_t\}_{t\geq 0}$ and predictable. Informally, $\lambda_t = E[dQ_t|\mathcal{F}_{t-}]$ and can be understood as the propensity for an initiating event to arrive in the next instant of time given the history of initiating events and reactor protections.

Now, consider the martingale transform M_t^C of protection unavailability $(1 - A_t)$ with respect to M_t of Eq. (41), where $M_t^C \stackrel{\text{def}}{=} \int_{[0,t]} (1 - A_u) dM_u = \int_{[0,t]} (1 - A_u) dQ_u - \int_{[0,t]} (1 - A_u) d\Lambda_u$.

Proposition 1 (Core Damage Martingale) $\mathbf{M^C} \triangleq \{M_t^C; t \geq 0\}$ is a martingale whenever the stochastic intensity process of arriving initiating events $\{\lambda_t\}_{t\geq 0}$ exists.

Proof: Since A_t is $\mathcal{F}_t$–predictable and $0 \leq A_t(\omega) \leq 1$ for all $\omega \in \Omega$ and $t \geq 0$, it follows that $\{M_t^C\}_{t\geq 0}$ is also an $\mathcal{F}_t$–martingale (see [17]), and noting that $Q_t^C = \int_{[0,t]} (1 - A_u) dQ_u$ counts the number of core damage events in the interval $[0, t]$, we have that

$$M_t^C = Q_t^C - \int_{[0,t]} (1 - A_u) d\Lambda_u \tag{45}$$

and, substituting from Eq. (43) gives

$$M_t^C = Q_t^C - \Lambda_t^C. \tag{46}$$

We refer to $\mathbf{M^C}$ as the *Core Damage Martingale* and its compensator is given by

$$\Lambda_t^C = \int_{[0,t]} \lambda_u^C du = \int_{[0,t]} (1 - A_u)\, \lambda_u\, du. \tag{47}$$

Remark 8 Eq. (46) stands as the most general available expression characterizing the relationship among core damage events, the arrival of initiating events, and the efficacy of reactor protections. It is important to keep in mind that, for all $t \geq 0, M_t^C, Q_t^C, \lambda_t^C, A_t$, and (in particular) λ_t are all random variables. Hence, Eq. (46) is nontrivial and must be examined in the context of stochastic integration.

Consider now, the following proposition:

Proposition 2 (Existence of CDF) If

$$\lim_{t\to\infty} \frac{M_t^C}{t} \overset{\text{a.s.}}{=} 0, \tag{48}$$

then λ^C exists (and is possibly a random variable) and for almost (a.a.) $\omega \in \Omega$

$$\lambda^C(\omega) = \lim_{t\to\infty} \frac{Q_t^C(\omega)}{t} = \lim_{t\to\infty} \frac{1}{t} \int_{[0,t]} (1 - A_u(\omega)) \lambda_u(\omega) du \tag{49}$$

where, $0 < \lambda^C(\omega) < \infty$. That is,

$$\begin{aligned} &\lim_{t\to\infty} \frac{M_t^C}{t} \overset{\text{a.s.}}{=} 0 \;\text{ if and only if} \\ &\lim_{t\to\infty} \frac{Q_t^C}{t} \overset{\text{a.s.}}{=} \lambda^C \;\text{ and} \\ &\lim_{t\to\infty} \frac{1}{t} \int_{[0,t]} (1 - A_u) \lambda_u du \overset{\text{a.s.}}{=} \lambda^C. \end{aligned} \tag{50}$$

Proof: Proposition 2 is an obvious consequence of Definition 5.1 and Eq. (46).

Remark 9 Clearly, CDF exists only if

$$\frac{M_t^C}{t} \overset{\text{a.s.}}{\to} 0 \text{ and } \lambda^C \overset{\text{a.s.}}{=} E\left[\lambda^C\right] < \infty. \tag{51}$$

Proposition 2 reveals the challenge in estimating CDF. In the absence of observed core damage events, predictive estimates of CDF must be formulated in terms of phenomena that can be observed. To this end, analysts must rely on observations of initiating event arrival times, and reactor protection performance (principally in the form of maintenance records and failure data). These observations are, of course, insufficient to capture the joint dynamics of $\{(A_t, \lambda_t); t \geq 0\}$ needed to directly employ the strong law relationship of Proposition 2, where,

$$\lambda^C \overset{\text{a.s.}}{=} \lim_{t\to\infty} \frac{\Lambda_t^C}{t} = \lim_{t\to\infty} \frac{1}{t}\int_{[0,t]} (1 - A_u)\lambda_u du. \tag{52}$$

Monte Carlo methods do not escape the difficulty of computing λ^C. **A** and $\{\lambda; t \geq 0\}$ are not mutually independent (even when **Q** is Poisson with rate λ) and λ_t is not directly observable. Since this dependence requires a Monte Carlo model to rely on an accurate characterization of the probability law on the joint stochastic process $\{(A_t, \lambda_t); t \geq 0\}$, it is clear that the data requirements to support accurate estimation of this probability law are beyond the practical reality of reactor unit operations records.

5.4 Cumulative assumption: arriving initiating events see time averages

Important insights regarding CDF are revealed by exploring the expectation of $\mathbf{M}^\mathbf{C}$. In particular, we are interested in the consequences of the stochastic dependence the state of system protections $\{A_t\}_{t\geq 0}$ and the arrival of initiating events $\{Q_t\}_{t\geq 0}$ and consequently $\{\lambda_t\}_{t\geq 0}$. Consider now,

Proposition 3 (Moment Convergence) Suppose that

$$\frac{M_t^C}{t} \overset{\text{a.s.}}{=} 0. \tag{53}$$

Then,

$$\begin{aligned} &\lim_{t\to\infty} E\left[\frac{M_t^C}{t}\right] = 0 \text{ if and only if} \\ &\lim_{t\to\infty} E\left[\frac{Q_t^C}{t}\right] = E\left[\lambda^C\right] \text{ and} \\ &\lim_{t\to\infty} E\left[\frac{1}{t}\int_{[0,t]} (1 - A_u)\lambda_u du\right] = E\left[\lambda^C\right]. \end{aligned} \tag{54}$$

And, with the additional condition that $\lambda^C = E\left[\lambda^C\right] < \infty$,

$$\text{CDF} = \lim_{t\to\infty} \frac{1}{t}\int_{[0,t]} E[(1 - A_u)\lambda_u]du. \tag{55}$$

Proof: Recall that almost sure convergence implies convergence in expectation (see [18]). Hence, Eq. (54) follows directly from Proposition 2. Nonnegativity of the integrand in Eq. (54) allows a routine application of Tonelli's Theorem to exchange the order of expectation and integration to show Eq. (55) (see [19]). Finally, $\lambda^C = E\left[\lambda^C\right] < \infty$ implies that

$$\frac{Q_t^C}{t} \overset{\text{a.s.}}{\to} E\left[\lambda^C\right] < \infty, \tag{56}$$

thus, ensuring the existence of CDF.

Recalling the definition of covariance, it immediately follows that.

Corollary 1 When $\lambda^C = E\left[\lambda^C\right] < \infty$, then

$$\text{CDF} \overset{\text{a.s.}}{=} \lim_{t\to\infty} \frac{1}{t}\int_{[0,t]} cov((1-A_u),\lambda_u)du + \lim_{t\to\infty} \frac{1}{t}\int_{[0,t]} E\left[(1-A_u)\right]E\left[\lambda_u\right]du. \tag{57}$$

Proof: Simply apply the definitions of CDF and covariance.

Corollary 2 When $\lambda^C = E\left[\lambda^C\right] < \infty$,

$$\text{CDF} \overset{\text{a.s.}}{=} \lim_{t\to\infty} \frac{1}{t}\int_{[0,t]} E[(1-A_u)]E[\lambda_u]du \tag{58}$$

if and only if $E[\lambda_t|\mathcal{F}_t] = E[\lambda_t]$ for all $t \geq 0$.

Proof: It need only be shown that $cov((1-A_t),\lambda_t) = 0$ if and only if $E[\lambda_t|\mathcal{F}_t] = E[\lambda_t]$. First, assume that $E[\lambda_t|\mathcal{F}_t] = E[\lambda_t]$ for all $t \geq 0$ for all $t \geq 0$. Note that

$$\begin{aligned} E[(1-A_t)\lambda_t] &= E[E[(1-A_t)\lambda_t|\mathcal{F}_t] \\ &= E[E[(1-A_t)E[\lambda_t]|\mathcal{F}_t] = E[\lambda_t]E[E[(1-A_t)|\mathcal{F}_t] = E[\lambda_t]E[(1-A_t)]. \end{aligned} \tag{59}$$

Thus, it follows from the definition of covariance that $cov((1-A_t),\lambda_t) = 0$. With $cov((1-A_t),\lambda_t) = 0$ it follows trivially that

$$E[(1-A_t)\lambda_t] = E[\lambda_t]E[(1-A_t)]. \tag{60}$$

Corollary 3. When, $E[\lambda_t|\mathcal{F}_t] = E[\lambda_t] = \lambda < \infty$ and $A_t \overset{\text{a.s.}}{\to} A$, then

$$\text{CDF} \overset{\text{a.s.}}{=} \lambda \lim_{t\to\infty} \frac{1}{t}\int_{[0,t]} (1-A_u)du. \tag{61}$$

Proof: It follows from Eq. (58) that when $E[\lambda_t|\mathcal{F}_t] = E[\lambda_t] = \lambda < \infty$,

$$\text{CDF} \overset{\text{a.s.}}{=} \lambda \lim_{t\to\infty} \frac{1}{t}\int_{[0,t]} E[(1-A_u)]du, \tag{62}$$

and when $A_t \overset{\text{a.s.}}{\to} A$,

$$\lambda \lim_{t\to\infty} \frac{1}{t}\int_{[0,t]} (1-A_u)du \overset{\text{a.s.}}{=} \lambda \lim_{t\to\infty} \frac{1}{t}\int_{[0,t]} E[(1-A_u)]du, \tag{63}$$

Eq. (61) follows.

Remark 10 It is important to appreciate that Corollary 3 *does not* imply that the state of reactor protections **A** is independent of initiating event arrivals **Q**. The condition $E\left[\lambda_t|\mathcal{F}_t\right] = \lambda$ is simply congruent with the lack of anticipation property of A_t.[22] The conditions establishing Corollary 3 allow for the possibility that initiating events can cause a failure of system protections (in addition to the possibility protections were already failed immediately prior to arrival).

---

[22] Recall from Section 2 that the state of system protections A_t at time t, does not influence the arrival times of future initiating events.

Recall Eq. (38). Perhaps the most important consequence of assuming that arriving initiating events see time averages is:

Corollary 4. When $E[\lambda_t|\mathcal{F}_t] = \lambda$, then

$$\boxed{p \triangleq \lim_{t\to\infty} \frac{Q_t^C}{Q_t} = E[A].} \tag{64}$$

Proof: Substitute Eq. (61) into Eq. (39).

5.5 Cumulative assumption: initiating events form a Poisson process

Eq. (61) admits, as a special case, the condition that the initiating event counting process **Q** forms an ordinary Poisson process of rate λ. By the Watanabe characterization, it is well known that the stochastic intensity of a counting process is a fixed constant if and only if that counting process forms an ordinary Poisson process. Thus, when **Q** forms an ordinary Poisson process of rate λ, its intensity $\lambda_t = \lambda$ for all $t \geq 0$, where λ is a positive valued constant. Returning to Eq. (46), it follows that the constant λ can be moved outside the integral, giving

$$\Lambda_t^C = \int_{[0,t]} (1 - A_u)\lambda_u du = \lambda \int_{[0,t]} (1 - A_u) du. \tag{65}$$

With the elements of the compensator of the core damage martingale a.s. constant for all $t \geq 0$ as given by Eq. (65), a special case of Corollary 3 is created. The condition $E[\lambda_t|\mathcal{F}_t] = \lambda$ given in Corollary 3 is more general (and difficult to calibrate with historical data) than under the Poisson initiating event assumption (see [20]).

Remark 11 The Poisson initiating event assumption does *not* imply independence between **A** and **Q**; on the contrary, **A** and **Q** remain dependent since there is the possibility that arriving initiating events will disrupt protections and lead to an accident. Wolff gives a full development of *Poisson arrivals see time averages,* (see [14]).

An obvious benefit gained when **Q** is assumed Poisson is that statistical calibration of λ becomes accessible using historical data. Since all Poisson processes are stationary and ergodic, the parameter λ is easily estimated since

$$\lambda = \lim_{t\to\infty} \frac{1}{t} Q_t(\omega) \tag{66}$$

for a.a. $\omega \in \Omega$.

Assuming that **Q** is Poisson does not change Eq. (61). But, all Poisson processes implicitly carry the very strong independent increments requirement, meaning here that the number of initiating events appearing in any collection of disjoint time intervals are mutually independent. In practice, the independent increments condition is an extremely difficult to justify.

5.6 Cumulative assumption: the protection process is stationary and ergodic

Returning to Corollary 3 with $A = \lim_{t\to\infty} A_t$, it is clear that for a.a. $\omega \in \Omega$ the split fraction p is given by

$$p = E[A] = \lim_{t\to\infty} \frac{1}{t}\int_{[0,t]} (1 - A_u(\omega))du. \tag{67}$$

Calibrating p with historical data, as is preferred with PRA, is generally inaccessible in the limit. However, assuming that the protection availability process **A** is both stationary and ergodic guarantees that A is measurable with respect to its tail σ-algebra for all $t \geq 0$, and thus

$$p = E[A] = E[A_t] = P(A_t = 0), \tag{68}$$

for any fixed $t > 0$. Clearly, the best available estimate of the split fraction p along the (only) historically available ω is given by

$$p \approx \frac{1}{t}\int_{[0,t]} (1 - A_u(\omega))du. \tag{69}$$

Typically, practitioners rationalize stationarity and ergodicity of **A** by claiming knowledge of a *long history* of protection operations augmented with an extensive observed history of initiating events and equipment maintenance activity. Based on a experience they will designate where to establish time $t = 0$, in Eq. (69). However, the approximation Eq. (69) suffers a bias that is a consequence of well designed protections: The speed of convergence of Eq. (67) depends on observing initiating events that will cause protections to fail. Obviously, well designed reactor protections will experience few (if any) such events. This implies that even under the assumption that **A** is stationary and ergodic the Eq. (69) will be positively biased for all $t < \infty$.

5.7 Cumulative assumption: protection unavailability is independent of all initiating events

The optimistic bias of Eq. (69) disappears under the additional assumption that **A** and **Q** are independent stochastic processes. Under this circumstance, since both the initiating event process **T** and the protection availability process **A** are adapted to the filtration $\{\mathcal{F}_t\}_{t\geq 0}$ as described in Section 2, it follows that

$$E[A_{T_n}] = E[A_t] \tag{70}$$

for all $t \geq 0$ and $n \in \mathbb{N}$. Independence of **A** and **Q** is congruent with the belief that reactor protections are so robust as to never fail due to the impact of an initiating event. Or, equivalently, an initiating event will induce an accident only if it finds protections already out of service upon arrival.

Of course, the model assumption that **A** and **Q** are independent stochastic processes is very strong and hardly justifiable in practice. That said, PRA methodology relies directly on the cumulative assumptions leading to Eq. (69). The assignment of the split fraction value p appeals to *ad hoc* analysis (e.g., penetration factor) where little or no observed historical calibration data can be found.

6. Summary: reasonableness assumptions and the consequences of making them

Numbers such as CDF or Large Early Release Frequency (LERF) are intended to inform engineers, regulators, and other citizens on the quantitative level of risk posed by a commercial nuclear power plant. Risk in this context is against the Atomic Energy Act of 1954 as amended (AEA) requirement for "adequate protection" of the health and safety of the public.[23] Protection contemplated in the AEA is to help prevent harm to the public from uncontrolled release of radioactive material. Such harms can come from accident scenarios involving reactivity control and loss of core cooling following reactor shutdown. Protective systems are put in place under regulation to minimize the likelihood of such scenarios.

Sections 3–5 review counting of consequential outcomes in hazardous processes such as commercial nuclear power plant technologies as they relate to "unknown-unknowns" (Section 4) and the assumptions required to obtain a numerical result where data are sparse (Section 5), particularly with respect to CDF. In the review, it is shown that PQ on robust protective systems, where little or no data are available for calibration and validation, will produce optimistic estimates for risk of protection breakdowns leading to disastrous consequences. The progressive introduction of *assumptions required to obtain quantitative levels of risk* and their relationship to nuclear power plant processes in current popular methods based on PQ include: (a) there are no unknown-unknowns, (b) there are no absorbing states, (c) CDF is well-defined, (d) arriving initiating events see the time average of protection availability, (e) arriving initiating events form a Poisson process, (f) the protection availability process is stationary and ergodic, and (g) protection unavailability is independent of all initiating events. The implications of such assumptions, when adopted, is that the number obtained will underestimate the frequency of an accident by an unknowable amount.

Commercial nuclear power, as currently regulated by Western standards, is arguably the most safe energy source (see again [7], on "safe" and "safety") of all other energy technologies currently available. Regulatory standards, regulatory inspection and enforcement, management oversight, and engineering practice implement available engineering solutions against "unknown unknowns" that include safety margins, defense in depth, root cause analysis, and corrective action to mitigate the consequence of "unknown unknown" events as they actually appear or are imagined to appear. Investors, regulators, design engineers, operators, and the public can best review efficacy of protection using PQ up to the point of risk quantification. Such fundamental engineering processes as design, testing, maintenance, operation, and design revision are supported by PQ that produces and holds out hope for categorizing breakdown scenarios for example, by the level of support they afford. Risk management in modern fission reactor technologies is best managed when the acquisition of knowledge from observations during ongoing operations is applied to protection going forward. This implies a strong organizational commitment to efficacious root cause analysis and corrective action throughout an asset's lifetime.

[23] The adequate protection language appears for example in Sections 182 and 189 of the Atomic Energy Act.

Author details

Ernie Kee[1]* and Martin Wortman[2]

1 University of Illinois at Urbana-Champaign, Urbana, Illinois, USA

2 HazTechRisk.org, Galveston, Texas, USA

*Address all correspondence to: erniekee@illinois.edu

References

[1] Kemeny JG. The need for change, the legacy of TMI: Report of the President's Commission on the Accident at Three Mile Island, John G. Kemeny, chairman. Washington, D.C.: s.n.: for sale by the Supt. of Docs., U.S. Govt. Print. Off; 1979

[2] Rogovin M. Three Mile Island: A Report to the Commissioners and to the Public. Vol. 1250. Nuclear Regulatory Commission, Special Inquiry Group; 1980

[3] Tobias A. Decay heat. Progress in Nuclear Energy. 1980;**5**:1-93

[4] Solberg Ø, Njå O. Reflections on the ontological status of risk. Journal of Risk Research. 2012;**15**(9):1201-1215

[5] Cardano G, Wilks S. The Book on Games of Chance: "Liber de Ludo Aleae", Translated by Sydneu Henry Gould. New York, NY, USA: Holt, Rinehart & Winston; 1961

[6] Bernoulli J, Jacobi Bernoulli... Ars Conjectandi, Opus Posthumum. Accedit Tractatus de Seriebus Infinitis, et Epistola Gallice scripta de Ludo pilae reticularis. Basileae, Impensis Thurnisiorum, Fratrum – Werke 3, 1713. pp. 107-286

[7] Hansson SO. Safety is an inherently inconsistent concept. Safety Science. 2012;**50**(7):1522-1527

[8] Doorn N, Hansson S. Should probabilistic design replace safety factors? Philosophy & Technology. 2011; **24**(2):151-168

[9] Möller N, Hansson SO, Peterson M. Safety is more than the antonym of risk. Journal of Applied Philosophy. 2006; **23**(4):419-432

[10] Hansson SO. The epistemology of technological risk. Techné. 2005;**9**(2): 68-80

[11] McTaggart JE. The unreality of time. Mind. 1908;**17**(68):457-474

[12] Kolmogorov AN. Foundations of the Theory of Probability: Second English Edition. New York, NY, USA: Courier Dover Publications; 2018

[13] Øksendal B. Stochastic Differential Equations: An Introduction with Applications. Berlin, Germany: Springer Science & Business Media; 2013

[14] Wolff RW. Poisson arrivals see time averages. Operations Research. 1982; **30**(2):223-231

[15] Bremaud P. Point Processes and Queues: Martingale Dynamics. New York: Springer-Verlag; 1981

[16] Çınlar E. Probability and Stochastics. New York, NY, USA: Springer; 2011

[17] Rogers LCG, Williams D. Diffusions, Markov Processes and Martingales. 9th ed. Cambridge, UK: Cambridge University Press; 2000

[18] Dudley RM. Real Analysis and Probability. Boca Raton, FL, USA: CRC Press; 2018

[19] Folland GB. Real Analysis: Modern Techniques and their Applications. Vol. 40. New York, NY, USA: John Wiley & Sons; 1999

[20] Melamed B, Whitt W. On arrivals that see time averages: A martingale approach. Journal of Applied Probability. 1990;**27**(2):376-384

Chapter 6

Possible Applications of Modern Aqueous Homogeneous Reactors

Ahmed Shaker

Abstract

This chapter describes the potential of the aqueous homogeneous reactor, briefing readers on the physics and history of the subject, whilst providing both current and possible future applications for this reactor technology. These reactors were some of the first nuclear reactors ever constructed, and provided valuable information on critical mass and other nuclear physical properties on fissile solutions. The compact nature of these reactors, combined with their inherent safety characteristics, have made them attractive for the generation of medical radioisotopes and neutrons for experimentation. However, material corrosion issues and advanced development of solid-fuelled light water reactors would curtail much interest in the technology in the 50's. Although operating temperatures of this type of reactor are usually low, even such low temperature heat is useful in process and industry; such a reactor can be used for environmentally-friendly district heating or the supply of process heat in industry, and could even be used to produce hydrogen. With modern advances in physics and chemistry, and disruptions in conventional energy sources; such reactors in their modern form may serve an important role: supplying various energy demands that could be derived from nuclear power, but may not require more advanced and costly reactor technologies.

Keywords: nuclear engineering, nuclear power, nuclear energy, process heat, district heating, nuclear heating, nuclear desalination, desalination, hydrogen production, clean hydrogen

1. Introduction

As we face an increasing issue of energy scarcity in light of a global energy crisis [1], nuclear energy is once again gaining attention as a way to reduce our reliance on fossil fuels and the often questionable sources that supply them [2]. Interest in nuclear energy is some of the highest it has ever been, and has seen many nations such as Bangladesh and Egypt build nuclear power plants for the first time, or others like Japan, return to the construction of new reactors after a long hiatus on construction [3].

However, research and development in nuclear energy can be said to be progressing at a slower pace relative to other technologies. Many kinds of fission reactors have been developed since the first was assembled under university football bleachers in 1942, but only a few types are in commercial use today; most using solid fuel with some form of water as coolant and moderator. These solid-fueled reactors have performed well in their service as power generators, proving themselves through tens of thousands of reactor-hours of operational experience, but are often large, complex machines that have become increasingly difficult to finance and construct and are rather inflexible in their siting and energy output [4].

It is important to note that not every application of nuclear energy should be handled by a select few or even a singular reactor type(s), nor is it possible to do so in many cases. As the uses of nuclear energy expand, the flexibility in design of nuclear reactors should also increase. For example, high-temperature reactors are being developed to replace coal-fired boilers *in-situ* [5], and fast-spectrum reactors are being explored as a way to burn up plutonium and other minor actinides as part of closed-cycle nuclear waste management programs [6].

Amongst the flurry of hypothetical fission reactor designs that were being considered for development in the backdrop of the Manhattan Project, there is one that holds great promise for modern applications — but has been side-lined and underdeveloped. In 1944, the brilliant Enrico Fermi decided to construct the first of a set of reactors at Los Alamos that would use liquid fuel mixed homogeneously with water as the moderator (wartime secrecy stipulated that they be called "water boilers") [7]; it was a little spherical reactor, contained in a stainless steel vessel no more than a foot across, filled with a solution of uranyl sulfate and light water, reflected by beryllium and graphite, and called LOPO (for low power) as it produced almost no energy in operation.

LOPO reached criticality in May of 1944 and was instrumental in determining the critical mass of uranium solutions. As experiments with LOPO concluded, it was disassembled later that year to make way for a higher-power solution reactor: HYPO [7]. HYPO was a larger reactor with cooling provisions that allowed it to operate at a higher power of 5.5 kW and provide a stronger neutron flux for experiments. HYPO was brought critical in December of 1944 and would be the key for the neutronics measurements needed to design the nuclear fission assemblies for the first atomic weapons.

HYPO itself would be, in turn, upgraded to SUPO (**Figure 1**) in March of 1951 with improved cooling and an increased neutron flux in sustained operations up to 35 kW of power [7]. SUPO would go on to operate until 1974, and the whole water boiler program would provide critical information that would be useful for the construction of other aqueous homogeneous reactors to come.

After the success at Los Alamos, engineers at Oak Ridge set out in 1952 to construct a far more powerful solution reactor under the auspices of the Homogeneous Reactor Experiment (HRE) [8]. This reactor, the HRE-1 (**Figure 2**), was much larger than previously built, with a power level of 1 MW thermal. The HRE-1 had an output temperature of 250 degrees, and was also coupled to a steam turbine, demonstrating an ability to produce 140 kW of electric power from its heat (**Table 1**). The HRE-1 used a novel control system utilizing magnetically coupled neutron absorbing plates between its core and reflector, dispensing the need for traditional control rods.

The HRE-1 was the first solution reactor that was built with the goal of extracting power in mind and would go on to conduct experiments regarding control and power output. During such experiments, it was further confirmed that AHRs possessed a very strong negative thermal reactivity coefficient [8], owing to the effect of voiding in the solution at higher-than-designed power levels and subsequently reduced

Figure 1.
SUPO reactor vessel without its graphite reflector in place. Public domain.

moderation; something originally discovered during an accidental reactivity incursion with SUPO.

These aqueous homogeneous reactors (AHRs) usually have a simple construction relative to solid-fuelled reactors, often consisting of simple tank-like structures, and requiring far less fuel fabrication (a very expensive and complex task in itself) as the fuel is not held in rods, but kept in either a solution or a slurry [9]. This design makes the possibility of meltdown with reactors of this type highly unlikely, and reduces the volume of active coolant, possibly making containment structures more compact and economical.

They have found themselves today as convenient sources of neutrons and fission products, and are often used today to produce radioisotopes for medical and industrial uses [9], as processing of the liquid fuel solution can be done *on-line* to continuously extract target isotopes before they decay, or done after batches of short reactor runs. However, there are other characteristics of these reactors that could also make them attractive for a wider range of applications.

The thermal energy of the AHR may come in useful in applications where the complexity and cost of high-temperature reactors is not needed, such as for district heating and seawater desalination. With a high fuel burn-up and simple control stratagem, such reactors could be used for long periods of time, almost — if not entirely unattended, as clean and sustainable industrial and commercial heat sources, much like a traditional fuel-fired hot water heater or boiler. All the while, the aforementioned problem of hydrogen arising from radiolysis may actually be a blessing in an era where we are starting to use hydrogen as a fuel itself but find it difficult to produce without fossil fuels.

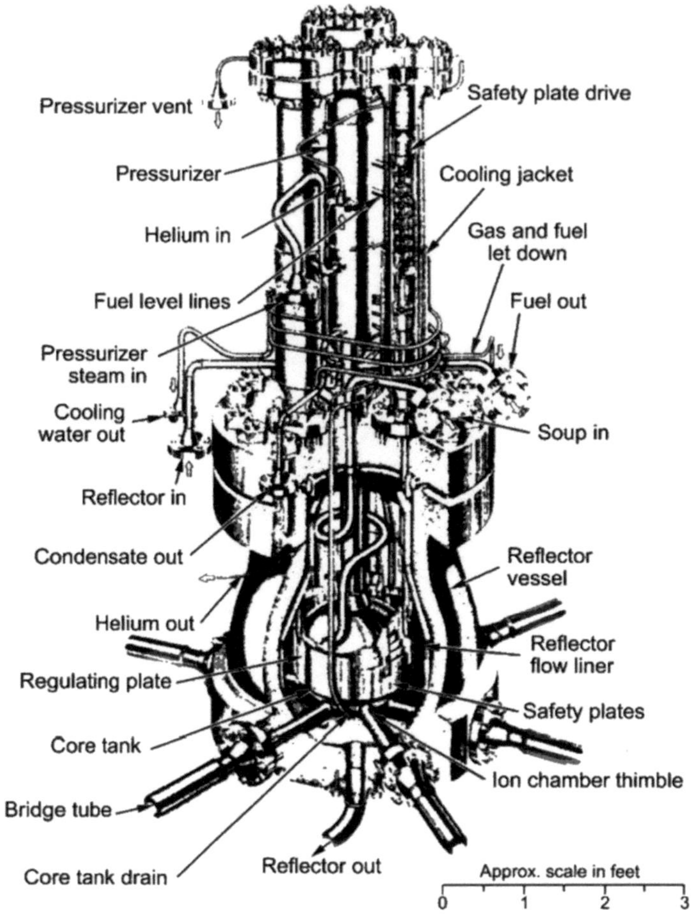

Figure 2.
HRE-1 internal details. ORNL drawing D-9065A. Public domain.

Thermal power	1000 kW (1600 kW max)
Fuel	93% enriched UO_2SO_4 in H_2O
Fuel inlet temperature	210°C
Fuel outlet temperature	250 °C
Core working pressure	6.9 MPa
Reflector	254 mm D_2O in pressure vessel, ⌀ 1067 mm, forged steel
Core vessel	⌀ 457 mm, type 347 stainless

Table 1.
HRE-1 design parameters.Adapted from Murray [8].

2. AHRs for the supply of heat

While the limited output temperature of the AHR makes it unsuitable for power generation, this low-grade heat is still usable for other industrial and commercial uses. A plethora of applications that currently utilize process steam could possibly be made nuclear using the AHR at a low capital cost, and replace combustion-fired boilers and heaters at many industrial sites.

The output temperatures of AHRs are limited partly by the high amounts of radiolysis encountered in operation; this, combined with heat, causes dissociation and precipitation of the solution at higher temperatures. These effects limit the output temperatures of AHRs to around 300°C, lower than the output temperature of modern PWRs and far lower than that of gas cooled designs, such as the 1200°C of the output of the HTR-PM [5].

However, with higher temperatures come higher pressures and stresses, requiring stronger, heavier, and more expensive structures to deal with such heat, especially those that hold water under pressure. As most AHR designs are not designed for two-phase flows (boiling) in the main reactor, this would necessitate a pressure vessel strong enough to keep the water from boiling at the reactor's operating temperature.

Such a pressure vessel could be selected for specific operating conditions and built accordingly, with lower temperature reactors requiring thinner vessel walls and lighter ancillaries compared to higher temperature reactors. This could make lower temperature reactors simpler and more economical to design and construct. Reactors could even be designed for near-atmospheric pressures if temperatures under 100°C are needed.

Low-complexity reactors that have been designed for such low-temperature applications are not a novel idea. For example, the SLOWPOKE Energy System was conceived by the AECL in Canada, and was a pool-type reactor with an output temperature of no more than 100°C. The low temperature allowed for non-pressurized construction and an increased margin of safety, obviating the need for operators; extensive automation was a goal of the design [10].

In the Czech Republic, the TEPLATOR project proposes to use spent PWR fuel bundles in a tank for the express supply of district heat at 98°C, and although it is not a homogenous design, and runs on spent fuel, the low capital cost of constructing such simple reactors may allow TEPLATOR to deliver heat at prices significantly lower than with fossil fuels [11]. A purpose-built AHR for such a purpose could, in theory, be able to do the same, if not more effectively.

Such heat from a low-cost and low-complexity source could enable greater use of nuclear power in seawater desalination and district heating. Contemporary multi-stage flash distillation (MSFD) processes can usually only work with temperatures under 120°C, while multiple effect distillation (MED) processes can only work with temperatures up to 70°C before scale formation becomes an issue [12]. Although nuclear desalination is a tried and tested concept, most implementations have used steam or hot water extracted from the turbines or condensers of power-producing reactors, increasing implementation costs. An AHR could be used solely as a heat source, and be designed for lower temperatures, potentially lowering implementation costs and making desalination a more accessible endeavor.

District heating has historically used network temperatures upwards of 200°C, but modern "fourth generation" systems are trying to bring this down to as low as 70°C [13] in a bid to capture more waste heat from various sources. As a compact and inherently safe design, AHRs could be sited very close to the built-up areas where district heating networks exist, lowering transmission losses and decreasing the need for expensive network piping. The heat could also be used to facilitate district cooling with the use of absorption chillers, allowing for such a plant to operate all-year round in temperate climates by supplying customers with chilled water.

Although direct production of steam is possible within the AHR itself (much like a BWR) [8] it is impractical due to the presence of fission products in the steam.

Because of this, most schemes to extract heat from AHRs would have to employ the use of intermediate heat exchangers and/or steam generators. Such a heat exchange loop would probably be of the *hi-lo-hi* type with high isolation loop pressure that is seen in nuclear desalination applications to preclude the possibility of contaminating network fluids with reactor solution [14].

3. AHRs for hydrogen production

The rapidly developing hydrogen economy currently relies on natural gas reforming to produce the vast majority of the substance, but has been transitioning to use sustainable sources of energy to lessen its reliance on fossil fuels. The use of hydrogen from nuclear sources would greatly help this effort. Most other schemes to produce hydrogen with nuclear energy involve either high temperature reactors using thermochemical cycles, or the use of electrolysis from nuclear electricity [15]. Such schemes involve equipment ancillary to the reactor itself, and require diversion of energy, reducing the power output of the associated plant in cogeneration. Such capital and running costs make nuclear hydrogen production difficult to viably implement at present.

However, the AHR is able to produce it *directly* as a by-product, potentially at a lower cost than other nuclear-based options, and would provide heat that could be useful for other parts of the hydrogen production process. As the very ions of uranium fuel in the AHR are in direct contact with the water as fission occurs, the intense energy from the fission, fission fragments and other energetic particles cause the water to disassociate to its constituents: hydrogen and oxygen [16]. This radiation-driven process is called radiolysis, and is responsible for the large amounts of hydrogen produced by the AHR. However, combined with the oxygen produced, this potentially explosive atmosphere would constitute a hazard inside the reactor and necessitates its constant recombination and removal.

Soluble ionic catalysts were proposed and tested to reduce this production, with copper shown to be effective in virtually stopping all production with the HRT [8]. But the hydrogen created by the reaction may be a far more valuable commodity going forward. World hydrogen demand is projected to grow 5 to 7 times larger than it was in 2021 by 2050, with hydrogen making up 15–20% of all energy demands [17].

As hydrogen is known to cause embrittlement issues in many materials, and creates an explosive atmosphere, the construction of such a reactor would require the use of novel techniques for extraction of the hydrogen and mitigation of the associated corrosion. Engineers at Los Alamos used mechanical gas separators that would use swirl-vanes to centrifuge out the gas in the HRT and send it to recombiners [18]. It should be noted, however, that hydrogen production in the AHR requires moderately high temperatures, and it would be inefficient to use AHRs to solely produce hydrogen [19]. Such hydrogen producing AHRs would inevitably have to be heating or power reactors, where the production of hydrogen could be seen as a bonus product or be used as a process gas alongside its associated nuclear process heat in industrial settings.

4. AHRs for the supply of power

As demonstrated with the HRE, production of power using steam turbines is possible with the AHR. By running the core solution through a steam generator, the HRE-1 was able to produce almost 1 MW of steam at 250°, giving its 140 kW power plant an efficiency of 14%. Later, the HRT increased this temperature to 300°, but had issues with the solution dissociating; creating hot spots and power fluctuations. These issues were caused by phase instabilities of U2SO4 in water at temperatures above 340° [8], and limit the efficiency of such a setup. Increasing the temperature beyond this also increases the corrosion of reactor internals greatly. Other experiments with different acid-based chemistries were able to increase the operating temperature to 450° [16], but high rates of corrosion make such exotic chemistries impractical even today.

Due to these corrosion issues limiting the output temperature, an AHR would not be the best choice of reactor for power production, but nonetheless, could be a simple way to implement nuclear power in mechanical or electrical applications. The design and construction advantages could be useful in applications where the increased fuel costs from the inefficiencies of the reactor plant do not warrant the need for a more efficient, yet, more capital intensive type of reactor. Such applications could include powering remote towns and facilities, or powering coastal and riverine vessels as part of a nuclear propulsion scheme.

An AHR could also be coupled to a stirling engine, eliminating the need for a traditional steam plant and associated steam generators, potentially reducing implementation costs in certain scenarios. Another option would involve using an organic rankine cycle (ORC), as temperatures below 350° make the use of steam rankine cycles particularly inefficient. ORC systems use an organic working fluid, such as pentane or R134a to achieve higher efficiencies at lower temperatures, and have seen use for capturing waste or low-temperature heat from industrial processes and renewable energies [20]. Such alternative cycles would make the power production from AHRs a more efficient undertaking in the present day.

Given its placid control characteristics, an AHR power plant could be operated autonomously, only needing personnel on-site for periodic maintenance and refueling. This characteristic, combined with the potential for cogeneration with heat, potentially makes such plants particularly attractive for arctic areas, where keeping personnel around is a challenge in itself.

If further drops in implementation costs are desired, the AHR could be used as a two-phase reactor, and include boiling in the core [16]. This would eliminate the need for steam generators, but would also expose the steam plant to fission products, contaminating it in the process. Such a design would also be problematic for the supply of heat, as it would have to be extracted from either the condenser, or off of the turbine stages, and then sent through a heat exchanger to render it safe for use.

5. AHRs as radioisotope sources

Owing to the compact size and intense neutron flux provided by the AHR, it makes it an excellent producer of radioisotopes from various sources [use the IAEA book on Mo-99 production for this section]. The solution chemistry of the AHR allows it to burn-up uranium almost completely and simultaneously remove fission products — both poisons and potential commodities, such as Xe-135 or Mo-99, for use as industrial and commercial sources of radiation.

The extraction of Mo-99 in particular from AHRs is of great interest. Mo-99 is an essential radioisotope for the medical industry, as it is used to produce Tc-99m for use in medical imaging as a tracer [21]. Mo-99 however, has a very short half-life of 66 h, which complicates its production and transportation. The use of an AHR would allow for the use of on-line processing to continuously produce Mo-99 whenever needed.

By use of chemical extraction methods, the Mo-99 condensate can be collected from the reactor loop without removing the fissile fuel. This method would also eliminate the amount of waste that would be generated when using a solid uranium target, of which the uranium is only 0.4% spent, and the processing of which releases fission products and unused uranium into waste streams [22].

Other isotopes like Sr-89 are often also difficult to produce using solid target-based systems. With the AHR, it is possible to create Sr-89 without the associated Sr-90 impurity by using extraction of the precursor gasses. At the ARGUS reactor at the Kurchatov Institute, Sr-89 of high purity is extracted by running the reactor for a few minutes at a time to generate the needed krypton precursors, then waiting for the shorter-lived Sr-90 precursor, Kr-90 (half-life of 33 s), to decay out faster than the relatively long-lived Sr-89 precursor, Kr-89 [22].

This Kr-89 is then pushed out of the reactor into a sorbent bed with the help of inert gasses, where it decays to produce Sr-89 and awaits further processing to remove other fission products. After purification, the Sr-90 content of the Sr. recovered is insignificant. Although running such a reactor for the sole purpose of extracting Sr-89 would be impractical, it is something that can be done alongside the production of Mo-99 and other radioisotopes.

The production of other useful isotopes such as Xe-133 — another useful medical tracer, I-131 — a radiotherapy agent, and Cs-137 — a potent gamma emitter with many industrial uses, is also possible, and was demonstrated with ARGUS [22]. If other radioisotopes needed to be produced in solid form, an AHR could be designed with provisions for the insertion of targets. If, for example, Co-60 needed to be produced in rod form, the reactor could be designed to accommodate for irradiation of such rods within its core.

Other radioisotopes of interest may be the Pu-239 or U-233 generated during fission by transmutation in uranium and thorium powered reactors. By using a breeder design with a "blanket" solution around the core, lost neutrons could be absorbed by fertile material such as U-238 or Th-232, rendering it fissile [16]. Such a system could also be adapted, in theory, to produce H-3 (tritium) from either H-2 or Li-6 for use in many applications.

6. Conclusions

The unique design features of the aqueous homogeneous reactor make them a valuable tool in increasing the uses of nuclear power. The simple design and control characteristics of AHRs could make them very useful wherever heat and/or steam is required, but the complexities of solid-fuelled reactors are unwarranted and unwanted. Moreso, if hydrogen needs to be produced, the AHR is capable of doing so directly. Various radioisotopes or fertile fuels could be extracted from such reactors as a value-added product during their normal operations, or be produced by reactors exclusively designed to do so.

Industrial and commercial users could benefit from a heat source that has the potential to be far cheaper than fossil fuels, yet cause far less environmental harm. This will be especially important in an era where governments worldwide are trying to wean themselves off of fossil fuels, which supply over 50% of the world's heating needs [23]. AHRs can supply heating to entire district heating networks, or be made small enough to supply individual buildings. Cooling can also be provided with the use of absorption chillers, and if done so while generating electricity, could comprise an economical CHP (combined heat and power) or CCHP (combined cooling, heat and power) scheme.

Thermal energy from the AHR could also be used for nuclear desalination purposes, where the energy costs of the process heavily determine the price of the purified water produced. Low-cost nuclear desalination could help arid areas of the world gain access to clean water without the associated fossil fuel needs or pollution. An AHR built expressly for this purpose has the potential to be far more economical than using power reactors or combustion boilers, as its material, logistic and safety requirements are greatly relaxed.

The production of hydrogen in the AHR is something that would usually comprise a problem, but with clever extraction techniques, this annoyance may turn out to be a lucrative opportunity in light of the growth in the hydrogen economy. Of course, if the complexities arising from hydrogen production in such a reactor are found to be a concern, catalysts may be added to the solution to facilitate complete recombination of these radiolysis products. But, as very little extra equipment is required for its extraction, the AHR could easily produce hydrogen as a value-added product for sale, or for use in industrial processes as feedstock or fuel. As the use of hydrogen as a fuel grows worldwide, the AHR finds itself uniquely suitable for the production of the gas without emissions or extra costs.

The production of radioisotopes would be another value-added product that could be produced in heating reactors, or produced by reactors purposely built for such a purpose. The solution chemistry of AHRs allow for the extraction of a wide range of radioisotopes without the production of large volumes of waste as seen with traditional target-based systems, and permits the extraction of certain radioisotopes that would otherwise be impractical to do with traditional systems, such as Sr-89. This same chemistry allows AHRs to achieve a high fuel burn-up, as well as a high breeding ratio, potentially allowing AHRs to operate as breeder reactors.

Although the output temperatures of AHRs are limited, they can still be used to generate power as demonstrated before (albeit, with lower efficiency than higher-temperature reactors). An AHR could provide just enough electricity through a Rankine or Stirling cycle to power itself and associated facilities, or generate enough to power other loads, such as industrial, commercial, and residential consumers of electricity. If done alongside thermal and hydrogen production, this could make the AHR a very flexible reactor type in cogeneration. This energy could also be used for propulsion at sea and inland waterways, where such reactors may enable economically-built nuclear-powered vessels that would require little justification for a more complex reactor type, such as tug boats or coastal bulk freighters.

As the world increasingly becomes more conscious of the deleterious effects of fossil fuel consumption and strives to move away from fossil fuels altogether, nuclear energy will be a vital tool in facilitating its replacement. In doing so, many different designs for nuclear reactors will be needed for the different applications they will be optimal for. The AHR has the potential to economically decarbonize industries and sectors that rely on low-temperature heat, or supply clean electric power. Hydrogen created from the regular operation of such reactors may be harnessed and utilized as a fuel, or be used for other industrial processes. These reactors are also optimal for the production of radioisotopes — especially those for medical uses, and could be made compact enough to be installed near or at the facilities that need them on a regular basis. Using modern technologies, it is not beyond the realm of possibility that the AHR can be deployed for commercial use in the near future. With its inherently safe and simple features bolstered by the use of our advanced tools and knowledge, the AHR is a very promising design looking forward.

Acknowledgements

Special thanks to Bianca, whose hospitality was instrumental in the creation of this chapter. I owe her my utmost gratitude.

Author details

Ahmed Shaker
University of Ontario Institute of Technology, Oshawa, ON, Canada

*Address all correspondence to: ahmed.shaker@uoit.net

References

[1] IEA. World Energy Outlook 2022 [Internet]. 2022. Available from: https://www.iea.org/reports/world-energy-outlook-2022 [Accessed: 29 January 2023]

[2] S&P Global. Nuclear energy is growing on a global scale [Internet]. 2022. Available from: https://www.spglobal.com/esg/s1/research-analysis/nuclear-energy-is-growing-on-a-global-scale.html [Accessed: 29 January 2023]

[3] Nikkei Asia. Japan PM Kishida orders new nuclear power plant construction [Internet]. 2022. Available from: http://asia.nikkei.com/Politics/Japan-PM-Kishida-orders-new-nuclear-power-plant-construction [Accessed: 29 January 2023]

[4] Bartak J, Bruna G, Cognet G. Economics of Small Modular Reactors: Will They Make Nuclear Power More Competitive? Journal of Energy and Power Engineering. 2021;**15**:193-201. DOI: 10.17265/1934-8975/2021.06.001

[5] Xu S, Lu YHM, Mutailipu M, Yan K, Zhang Y, Qvist S. Repowering coal power in China by nuclear energy—Implementation strategy and potential. Energies. 2022;**15**(3):1072. DOI: 10.3390/en15031072

[6] Daniel W. Why Faster Is Better - on Minor Actinide Transmutation in Hard Neutron Spectra [thesis]. Stockholm: KTU; 2007

[7] Bunker Merle E. Early reactors: From Fermi's water boiler to novel power prototypes. In: Los Alamos Science No. 7. Los Alamos: Los Alamos National Laboratory; 1983. pp. 124-131

[8] Murray WR. An Account of oak Ridge National Laboratory's Thirteen Nuclear Reactors. Oak Ridge: Oak Ridge National Laboratory; 2009. pp. 14-23

[9] Rijnsdorp S. Design of a Small Aqueous Homogeneous Reactor for Production of 99Mo [thesis]. Delft: Delft University of Technology; 2014

[10] Hilborn JW, Lynch GF. SLOWPOKE - Heating Reactors in the Urban Environment. Chalk River: Chalk River Nuclear Laboratories; 1988. pp. 1-6

[11] Hussein ASA, David M, Michael M, Radek S. Economics of reusing spent nuclear fuel by Teplator for district heating applications. International Journal of Energy Research. 2022;**46**(5):1. DOI: 10.1002/er.7521

[12] Alharthi K. Increasing the Top Brine Temperature of Multi-Effects Distillation (MED) to Boost its Performance through Controlling the Formation of Scale by Nanofiltration and Antiscalants. Thuwal: King Abdullah University of Science and Technology; 2020. p. 15

[13] Lund H, Werner S, Wiltshire R, Svendsen S, Thorsen JE, Hvelplund F, et al. 4th Generation District heating (4GDH) integrating smart thermal grids into future sustainable energy systems. Energy. 2014;**68**:5. DOI: 10.1016/j.energy.2014.02.089

[14] Bouaichaoui Y, Belkaid A, Amzert SA. Economic and safety aspects in nuclear seawater desalination. Procedia Engineering. 2012;**33**:151. DOI: 10.1016/j.proeng.2012.01.1187

[15] Non-Nuclear Products of Nuclear Energy. Issy-les-Moulineaux: OECD NEA; 2004. pp. 9-11

[16] Lane JA, Thompson WE. Homogeneous reactors and their development. In: Lane JA, MacPherson HG, Maslan F, editors. Aqueous Homogeneous Reactors. Reading: Addison-Wesley; 1958

[17] Energy Transitions Commission. Making the Hydrogen Economy Possible: Accelerating Clean Hydrogen in an Electrified Economy [Internet]. 2021. Available from: https://www.energy-transitions.org/publications/making-clean-hydrogen-possible/ [Accessed: 21 February 2023]

[18] Hafford JA. Development of the Pipe-Line Gas Separator. Oak Ridge: Oak Ridge National Laboratories; 1954

[19] Kerr W, Majumdar DP. Aqueous homogeneous reactor for hydrogen production. In: Proceedings of the Hydrogen Economy Miami Energy Conference; 18-20 March 1974; Miami Beach. New York: Plenum Press; 1975. pp. 167-181. DOI: 10.1007/978-1-4684-2607-6_12

[20] Rettig A, Lagler M, Lamare T, Li S, Mahadea V, McCallion S, et al. Application of organic Rankine cycles. In: Proceedings of the World Engineer's Convention; 4-9 September 2011; Geneva. ZHAW: Winterthur; 2011

[21] Papagiannopoulou D. Technetium-99m radiochemistry for pharmaceutical applications. Journal of Labelled Compounds and Radiopharmaceuticals. 2017;**60**(11):502-520. DOI: 10.1002/jlcr.3531

[22] Chuvilin DY, Khvostionov VE, Markovskij DV, Pavshouk VA, Zagryadsky VA. Low-waste and proliferation-free production of medical radioisotopes in solution and molten-salt reactors. In: Rahman RA, editor. Radioactive Waste. London: IntechOpen; 2012. p. 142, 153-160. DOI: 10.5772/37158

[23] IEA. Heating - Fuels & Technologies [Internet]. 2022. Available from: https://www.iea.org/fuels-and-technologies/heating [Accessed: 21 February 2023]

Section 4

Applications

Chapter 7

Applications of Fission

Anna C. Hayes

Abstract

This chapter is devoted to a discussion of applications of nuclear fission. It covers some aspects of the topics of nuclear reactors, nuclear safeguards and nonproliferation, reactor antineutrinos, and nuclear medicine. It is, however, limited in scope and the reader is encouraged to explore the many other exciting sub-areas of the applications of nuclear fission.

Keywords: nuclear reactors, MOX fuel, fission antineutrinos, nuclear nonproliferation, radio-nuclide therapy

1. Introduction

The two key properties of fission, the release of several neutrons and the release of a broad distribution of fission fragments, are fundamental reasons for the huge number of applications that have come since the discovery of the fission process. The importance of neutron emission mainly lies in the uses of fission chain reactions, while the radioactive nature of most of the fission fragments underlies applications from this aspect of fission. The chain reaction nature of fission can be exploited in two ways, controlled fission chain reactions used to generate nuclear energy and uncontrolled chain reactions in explosive nuclear devices. Both of these uses of fission have been in existence since the Second World War when many of the top scientists of the time concentrated on making these applications a reality. Although today the basic and practical principles determining our ability to use fission for nuclear energy or nuclear explosions are well-understood, the initial years required a lot of ingenuity and creativity. The research was equally creative, mostly started in the 1950s, that went into harnessing the products of fission for medical isotopes. Several fission fragments have ideal decay properties to be used as diagnostics or in radiotherapy to treat some medical conditions, especially cancer. The decay properties of many fission fragments also make them ideal probes for nuclear nonproliferation and reactor safeguards. Another property of the decay of fission fragments is that their beta-decay results in the emission of antineutrinos and the antineutrino fluxes from nuclear reactors have been used in several worldwide collaborations to examine neutrino oscillations and physics beyond the Standard Model of particle physics.

The enormous range and diversity of applications of nuclear fission make it one of the most intriguing subatomic processes. In this chapter, I will attempt to discuss a selection of these applications, ranging from nuclear reactors to nuclear nonproliferation to neutrino physics. It is not possible to cover all of the applications of fission that have been invented by so many resourceful scientific teams. But it is my hope that the discussion of the applications presented here will encourage the reader to explore this rich field in more depth.

2. Nuclear reactors

This section discusses aspects of nuclear reactors that are important for understanding later discussions of nonproliferation and reactor neutrino physics. In particular, this section pays particular attention to the production of plutonium in reactors. Most modern reactors use low-enriched (~2–5% ^{235}U) uranium fuel and thermalized neutron flux. The shape of the neutron flux depends on the initial enrichment, and in **Figure 1**, examples are shown for a pressurized water reactor (PWR) with different enrichments. These fluxes were derived from MCNP/CINDER [1] simulations. With fresh fuel, the burn is initially dominated by fissions from ^{235}U, but as the burn proceeds, ^{239}Pu is generated by neutron capture on ^{238}U, followed by two beta decays, **Figure 2**.

The in-growth rate of the different Pu isotopes depends on the reactor design and the initial fuel enrichment of the uranium fuel. For example, in **Figure 3**, we compare the percentage of fissions induced by the different isotopes in the fuel for two PWRs that only differ in the fuel enrichment, one with 2.7% and the other with 4.2% enriched uranium. The total exposure of the 4.2% enriched fuel is about twice that of the 2.7% enriched fuel and is 60 GW days per metric ton of uranium (GWd/MTU). These simulations were taken from Ref. [2] and from earlier coupled calculations with the ERPI/CELL and CINDER codes [3, 4]. Apart from the factor of two differences in total burn times, the two fission histories are very similar. Indeed, increasing the fuel enrichment induces a kind of "accordion" effect in which the x-axis is simply stretched.

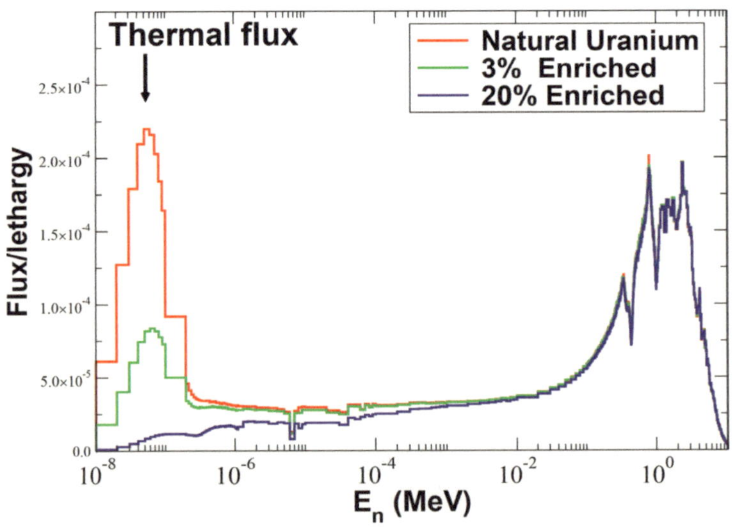

Figure 1.
The neutron flux of PWRs with different levels of ^{235}U enrichment. The flux is graphed as flux ($\Phi(E)$) per unit lethargy (Δu), that is, $\Phi(E)/\Delta u$, where $\Delta u = ln(E_I/E_{I+1})$.

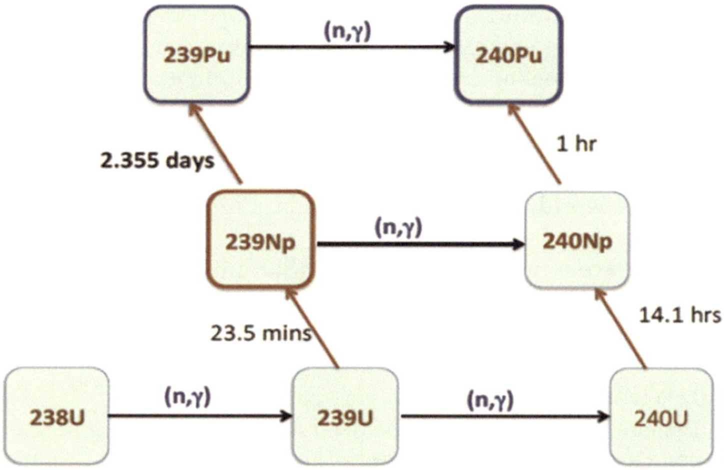

Figure 2.
Reactor production of ^{239}Pu and ^{240}Pu starts with neutron capture on ^{238}U. The production is complicated by the capture on and beta decay of the intermediate nuclei ^{239}U, ^{240}U, ^{239}Np, and ^{240}Np. At high neutron flux, $\varphi \sim 10^{14}$ n/cm^2/sec, the 2.355 half-life of ^{239}Np affects the ^{240}Pu/^{239}Pu ratio.

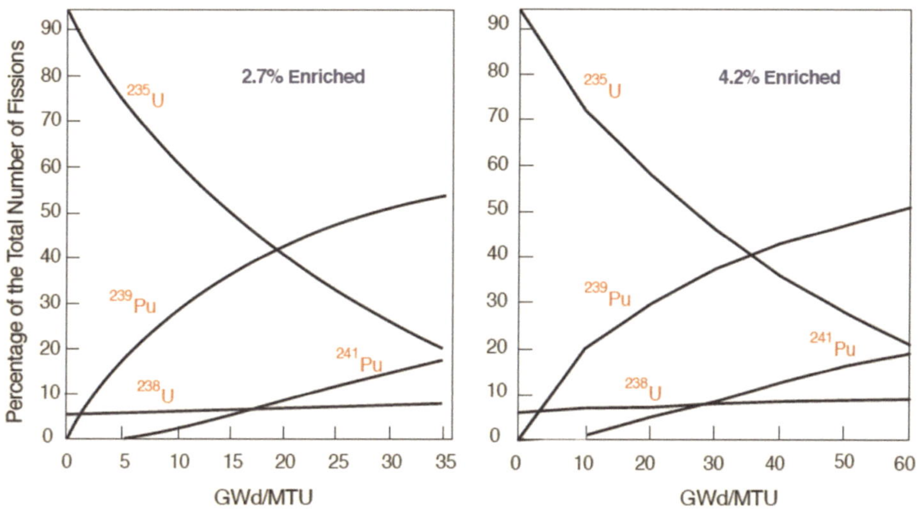

Figure 3.
The fission history for fresh 2.7% enriched (97.3% ^{238}U, 2.7% ^{235}U) and 4.2% enriched (95.8% ^{238}U, 4.2% ^{235}U) reactors. As the fuels burn, the importance of ^{235}U, in terms of the fraction of fissions taking place, decreases steadily. The fission histories are very similar for the two enrichments, apart from the clear difference in burn time.

2.1 Other reactor designs

A number of alternate reactor designs to PWRs exist or have been designed. Perhaps the most famous alternate design is the Canadian design, the CANDU reactor [5]. CANDU designs use natural uranium (0.711% ^{235}U, 9928% ^{238}U, and trace ^{234}U) and are pressurized heavy-water (D_2O) designs. Unlike conventional light-water PWRs, where the core fuel is placed in a pressure vessel, the CANDU bundles of fuel are contained in pressure tubes and these pressure tubes are then contained in a larger unpressurized vessel (a calandria) that is moderated by heavy water. In addition to being capable of producing energy from natural uranium, a large advantage of the CANDU design is that refueling can be done on a continuous basis. This is because only a signal fuel tube needs to be depressurized and replaced at a time. This is in strong contrast to a normal PWR, where the entire core must be shut down in order to open the pressure vessel and refuel the core. There are 30 CANDU reactors operating around the world, in Argentina, Canada, China, India, Pakistan, Romania, and South Korea.

Another very different reactor design is the thorium-uranium fuel design [6]. Thorium is more than three times more abundant on Earth than uranium and considerable research has gone into Th-U reactor designs. ^{232}Th is almost stable with a half-life of 1.405×10^{10} y, and it accounts for 99.98% of the natural thorium on Earth. The remaining 0.02% is accounted for by the very small abundance of ^{230Th} found in parts of the ocean. In general, ^{232}Th can only be used in breeder reactors because it does not undergo thermal fission. The breeding process is analogous to the production of ^{239}Pu in uranium reactor, in which ^{232}Th absorbs a thermal neutron to make ^{233}Th, and two beta decays result in ^{233}U, which is fissile.

$$^{232}Th + n \rightarrow {}^{233}Th(22.3 \text{ min}) \overset{\beta-decay}{\rightarrow} {}^{233}Pa\ (26.975 \text{ days}) \overset{\beta-decay}{\rightarrow} {}^{233}U \tag{1}$$

The thermal neutron capture cross section on ^{232}Th is larger than that on ^{238}U, which results in ^{233}U being bred more efficiently in thorium reactors than plutonium in uranium reactors. To sustain criticality in a thorium reactor, it normally requires a second fissile material to be included in the core to initiate the burn and to provide a neutron flux to allow the nuclear sequence in Eq. 1 to proceed. There have been many possible fissile fuels considered for adding to thorium-dominated fuel. These have mostly been low-enriched uranium (LEU) or reactor-grade plutonium (RGPu). LEU generally refers to uranium enriched up to 20% in ^{235}U. Higher enrichments would also work, but proliferation becomes an issue. RGPu is generally spent reactor fuel containing about 50% ^{239}Pu, with the remaining Pu coming from ^{238}Pu, ^{240}Pu, ^{241}Pu, and ^{242}Pu.

In **Figure 4**, we show the burn history for the case of 80%ThO_2 + 20%LEU. For this example, ^{235}U initially dominates the fissions, until sufficient ^{233}U has been bred from the ^{232}Th. However, the ^{238}U in the LEU fuel also produces 2^{39}Pu, which eventually contributes more fission than ^{235}U.

2.2 MOX plutonium fuels

One of the important concepts for the nuclear fuel cycle is to reuse spent reactor fuel by separating the plutonium and mixing it with depleted uranium to form a mixed oxide (MOX) fuel [7]. The plutonium is recovered through reprocessing. Depending on the nation involved in the reprocessing, the spent uranium can also be reprocessed. There are about 40 reactors in Europe (France, Germany Belgium, and Switzerland) that are licensed to use MOX, although only about 30 European reactors are currently using MOX fuel. Japan also recycles MOX fuel in its reactors. Typically, commercial thermal reactors in countries that recycle reactor fuel are loaded with one-third of MOX, although some advanced light water designs are capable of accepting 100% MOX loadings.

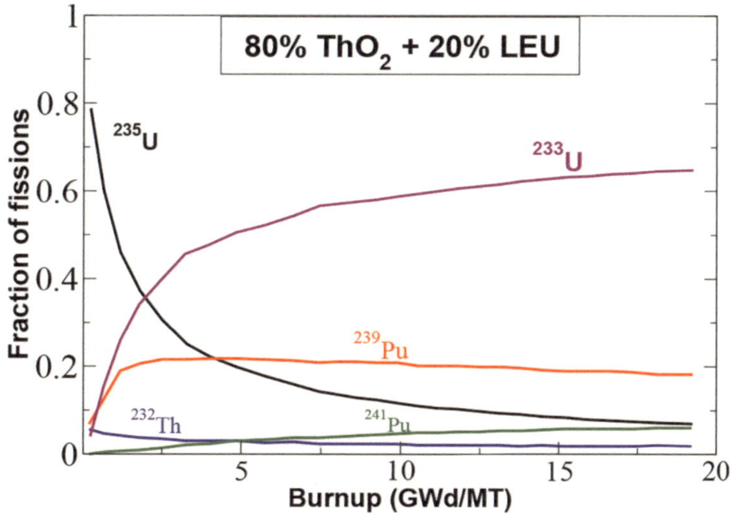

Figure 4.
The fission fractions predicted for a Th-U reactor as a function of the burnup. The main role of the ^{232}Th is to breed ^{233}U, and ^{233}U eventually dominates the number of fissions being produced. The LEU fuel is necessary to bring the reactor up to critical and supply enough neutrons for the breeding of ^{233}U. The LEU fuel is made up of 2.56% 235 U and 97.44% ^{238}U. The ^{239}Pu comes from neutron capture on ^{238}U, as summarized in ***Figure*** *2.*

Another concept that has been explored is to recycle weapons-grade MOX plutonium fuel. The typical isotopic difference between weapons-grade and reactor-grade MOX pluton is summarized in **Table 1**.

There are many MOX loadings that could be considered, but in a detailed analysis we considered [8] four possible cases:

- the entire reactor core is reactor-grade (RG) MOX fuel.
- the entire reactor core is weapons-grade (RG) MOX fuel.
- one-third of the core is RG MOX fuel; the remaining two-thirds of the fuel is assumed to be fresh 2.56% enriched LEU.
- one-third of the core is WG MOX fuel; the remaining two-thirds of the fuel is assumed to be fresh 2.56% enriched LEU.

For all four cases, the Monteburns code [1] was used, which couples the Monte Carlo neutron transport code MCNP [9] to the burn code CINDER'90 [10]; the latter uses 63 neutron energy groups and tracks up to 3400 nuclides, including 638 isomers. In all cases, we assume the same reactor configuration as the H.B. Robinson Unit 2 (HBR2) PWR, which was loaded with a fuel assembly that had 2.56% enriched fresh

Isotope %	235 U	238 U	238Pu	239Pu	240Pu	241Pu	242Pu
Weapons	0.679	93.62	—	5.0	3.0	—	—
Reactor	0.679	93.62	0.203	2.653	1.373	0.535	0.535

Table 1.
Initial isotopics for weapons and reactor-grade MOX fuels.

LEU and no MOX fuel. The simulations reproduce the reported history for the HBR2 PWR using variable concentrations of burnable boron poison rods over the four burn periods and the simulations predict spent fuel isotopic inventories for the uranium and plutonium isotopes within about 5% of measurement. The equivalent burn histories predicted for our four MOX fuel scenarios are shown in **Figures 5** and **6**.

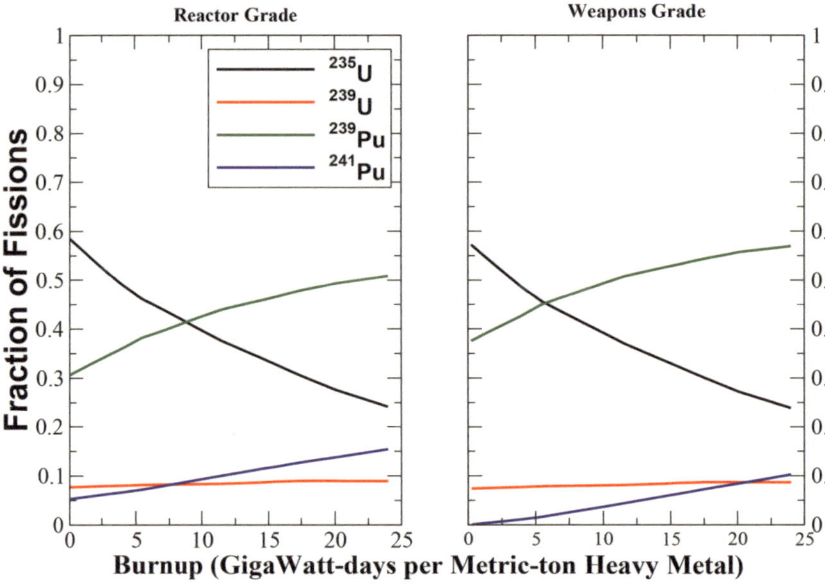

Figure 5.
Fission fraction for $^{235,238}U$, and $^{239,241}Pu$ as a function of burnup for 33.3% MOX and 66.7%LEU. Panel (a) is RG and (b) WG plutonium.

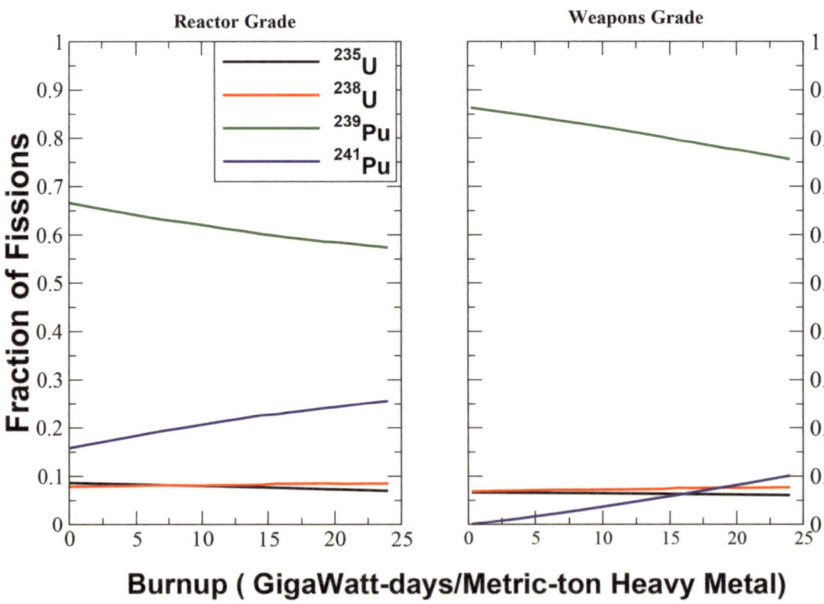

Figure 6.
Fission fraction for $^{235,238}U$, and $^{239,241}Pu$ as a function of burnup for pure MOX fuels. Panel (a) is for RG and (b) is for WG plutonium. The two grades of Pu are distinguished by the relative importance of ^{239}Pu and ^{241}Pu.

As expected, the burn histories for cores involving partial or total MOX fuel are quite distinct from fuel only containing LEU. In the cases where one-third of the core is MOX fuel, the fissions are dominated by ^{235}U and ^{239}Pu. The largest difference in the burn between the RG and WG plutonium fuel is the relative importance of ^{241}Pu. A second difference is the burnup value at which ^{239}Pu becomes a larger fraction of the fission than ^{235}U. As is usually the case with all uranium fuel, the relative importance of ^{238}U remains approximately constant throughout the burn. When the fuel is 100% MOX Pu, ^{239}Pu dominates the fissions for all burnups. Burning RG fuel differs from WG fuels in the relative importance of ^{239}Pu versus ^{241}Pu.

3. Fission reactor antineutrinos

One very important application of fission is the use of the antineutrinos emitted in the beta decay of fission fragments. Since the discovery of the antineutrino at the Savannah River reactor by Reines and Cowan in 1953 [11–13], neutrino physicists have taken advantage of the intense sources of low-energy antineutrinos from nuclear reactors to study the fundamental nature of neutrinos, particularly neutrino oscillations. The concept of neutrinos oscillating [14] from one flavor (ν_e, ν_μ, ν_τ) to another is a natural outcome of gauge theories with massive neutrinos and in its simplest form, it is expressed as a unitary transformation relating the flavor eigenstates (ν_e, ν_μ, ν_τ) to mass (ν_1, ν_2, ν_2) eigenstates.

Understanding the details of antineutrino fluxes from reactor is a complex and detailed subject. All of the detectable neutrinos emitted from reactors come from the beta-decay of neutron-rich fission fragments and correspond to an antineutrino spectrum ranging from 0 to 10 MeV. The antineutrino spectrum from each fissioning actinide is different and involves thousands of beta-decay branches. The study of neutrino oscillations from reactors has become increasingly important and has now reached a precision science, wherein detailed knowledge of the antineutrino spectra is required. This is especially true since the emergence of the so-called reactor neutrino anomaly [15].

The history of reactor neutrino experiments is a very rich one that spans over 60 years. These neutrino oscillation experiments mostly involved searches for variations of the emitted spectrum with the distance from the reactor. In the early experiments, performed in the 1980s–1990s, neutrino detectors were placed within 100 meters of the reactor, but no oscillations were observed. These observations agreed with predictions based on three-neutrino oscillations models and the findings of other neutrino experiments. The first observation of the disappearance of antineutrinos from a reactor was made by the KamLAND [16] experiment in the years between 2003 and 2008. KamLAND detected antineutrinos from 56 Japanese reactors at an average distance of about 180 km. By 2012, KamLAND, together with the CHOOZ [17, 18] and Palo Verde [19] experiments, had set an upper limit on the neutrino mixing angle θ_{13} and shown that θ_{13} was significantly smaller than the mixing angles θ_{12} and θ_{23}. Following these experiments, the reactor neutrino field moved to much larger detectors and employed both a near detector and a main far detector. The reactor neutrino experiment Daya Bay [20], RENO [21], and Double Chooz [22] determined the value of θ_{13} with high accuracy. The best-fit value from Daya Bay is $\sin^2(2\theta 13) = 0.084+/-0.005$.

3.1 The physics determining fission antineutrino spectra

For precision studies of neutrino oscillations, accurate knowledge is needed of the fission antineutrino spectra for each actinide contributing to the reactor fuel. This issue became a focal point of reactor neutrino studies when in 2011, a reevaluation of the spectra suggested by Huber [23, 24] and Mueller [25] (HM) stated that the expected spectra from all reactors should be increased by about 5–6%. If correct, this would introduce a very serious problem for the neutrino oscillations community because it would mean a need for reassessment of all previous reactor neutrino experiments. In particular, it would imply that the short-baseline experiments were observing neutrino oscillations, but this suggestion could not be accommodated by the standard three neutrino model. New models were proposed to explain the situation, including postulating the existence of a sterile neutrino.

A related problem arose from analyses of the change in the number of antineutrinos emitted from reactors with increasing fuel burnup [26]. As summarized in **Figure 3**, as the burn proceeds in a reactor, the relative importance of ^{239}Pu increases. The total number of antineutrinos emitted from the fission of ^{239}Pu is less than that emitted from ^{235}U, so as the fuel evolves the number of antineutrinos emitted decreases. Daya Bay [26] used the reactor fuel evolution to extract a cross-section ratio for ^{235}U/^{239}Pu, $\overline{\sigma}_{235}/\overline{\sigma}_{239}$ =1.445+/$-$0.097. Here $\overline{\sigma}$ means the average cross section for an antineutrino spectrum for the inverse beta-decay process $p + \overline{\nu}_e \rightarrow n + e^+$. However, the HM [23, 25] ratio is $\overline{\sigma}_{235}/\overline{\sigma}_{239}$ =1.53+/$-$0.05. The combination of the reactor neutrino anomaly and the anomalous $\overline{\sigma}_{235}/\overline{\sigma}_{239}$ ratio behooved the nuclear physics community to reexamine the detailed physics determining fission antineutrino spectra.

There are two complementary ways to determine the expected antineutrino spectrum from a fissioning nucleus [27]: the *ab initio* summation and the electron spectrum conversion methods. In either case, the aggregate fission antineutrino spectrum is determined by summing the contributions of all β-decay branches (b_{ni}) of all fission fragments ($Y_n(A_n, Z_n)$),

$$\frac{dN}{dE_{\overline{\nu}}} = \sum_n Y_n(Z_n, A_n) \sum_i b_{ni}\left(E_0^i\right) S_{\overline{\nu}}\left(E_{\overline{\nu}}, E_o^i, Z_n\right). \tag{2}$$

Here $Y_n(A_n, Z_n)$ is the cumulative fission yield for fragments (A_n, Z_n). The beta-decay spectrum S for a single transition in nucleus (Z, A) with end-point energy E0 = Ee + Eν is

$$S(E_e, Z, A) = S_0(E_e) F(E_e, Z, A) C(E_e)(1 + \delta(E_e, A, Z)), \tag{3}$$

where $S_0 = G_F^2 E_e (E_0 - E_e)^2/2\pi^3$, $E_e(p_e)$ is the total electron energy (momentum), $F(E_e, Z, A)$ is the Fermi function, and $C(E_e)$ is a shape factor [28] for forbidden transitions. In the case of Gamow-Teller transitions $C(E_e) = 1$. The corrections to the spectrum, including in the term $\delta(E_e, Z, A)$, arise from weak magnetism, finite size effects in the Fermi function, and radiative corrections. Changes in the treatment of these corrections were a major source of the initial reactor neutrino anomaly.

In the summation method, one uses the nuclear databases to carry out the sum of over all fission fragments and end-point energies, Eq. 2. This method suffers from the problem of the databases being somewhat incomplete. There are theoretical models for some of the missing data, but overall it would be difficult to use the summation method to determine whether an anomaly exists at the 5% level or not.

The second method for determining the antineutrino spectra is to convert a measured aggregate beta-spectrum into an antineutrino spectrum. This method also comes with difficulties. Any fit to an aggregate beta-spectrum is restricted to about 25–30 (fake) endpoint energies, E^i_0. This is because the measured [29–32] aggregate beta spectra are very smooth over five orders of magnitude, **Figure** 7. Thus, a prescription is needed for the 25–30 values of "Z_{eff}" that enter the Fermi functions and this introduces uncertainty at the few percent levels. However, regardless of the prescription used, it can be shown [33] that the ratio $\overline{\sigma}_{235}/\overline{\sigma}_{239}$ is constrained to be close to 1.53+/ −0.05, if the Schreckenbach data [29] are used for the conversion.

To address this problem, Kopeikin *et al.* [34] remeasured the ratio of the beta spectra for ^{235}U and ^{239}Pu at a research reactor at National Research Centre Kurchatov Institute. They found a ratio of $\frac{\overline{\sigma}_{235}}{\overline{\sigma}_{239}} = 1.45 + / - 0.03$, which is close to the average value found by Daya Bay and RENO, but 5% lower than any reasonable analysis of the Schreckenbach ^{235}U and ^{239}Pu aggregate beta spectra could predict. In addition, the STEREO experiment, which was performed at a reactor with highly enriched ^{235}U fuel, showed that the anomalous results for the change in the total antineutrino flux with fuel evolutions observed at Daya Bay and RENO could be explained entirely in terms of the Schreckenbach measurements having a ^{235}U spectrum with a normalization that was 5% too high. This then translated into the HM normalization also being too high. It is worth noting that the predictions of the summation method, using the

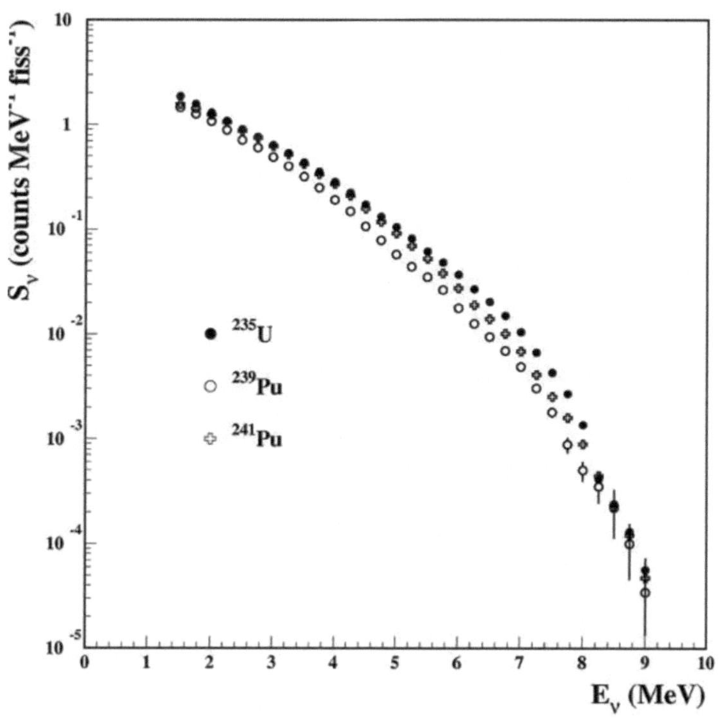

Figure 7.
The aggregate fission antineutrino spectra for ^{235}U, ^{239}Pu, and ^{241}Pu, deduced from the aggregate beta spectra measured by Schreckenbach et al. [29]. Conversion of these spectra to antineutrino spectra by Huber [23] and Mueller [25] led to the so-called reactor neutrino anomaly [15] and to the predicted ratio for ^{235}U/^{239}Pu antineutrinos inverse beta decay cross sections of 1.53+/−0.05, which is larger than that observed at Daya bay [26].

JEFF-3.1 cumulative fission yields [35] and the ENDF/B-VIII.0 decay data yields [36], yield a ratio $\frac{\bar{\sigma}_{235}}{\bar{\sigma}_{239}} = 1.445$, in agreement with the more modern experiments.

4. Nuclear threat reduction and global security

Preventing nuclear proliferation is a worldwide problem that relies on international agreements and on our ability to detect illicit nuclear activities. Undeclared production of fissile material at all stages of any such program is of most concern. These international efforts include nuclear safeguards for reactors, detecting reprocessing of spent nuclear fuel for the production of weapons plutonium, detecting trafficking of nuclear material, and developing techniques for analyzing debris in the case of a possible terrorist nuclear explosion. The techniques that can be used often depend on the standoff distance. For example, at a compliant reactor, measurements can be made on-site, while for a reprocessing facility in an unfriendly nation, the techniques used may require detection schemes that can work many kilometers from the facility. We consider here some very different situations.

4.1 Monitoring the fuel isotopic content in a molten salt reactor

The international community is paying increasing attention to molten salt reactor (MSR) technology as a promising form of clean and reliable energy. In addition to the fuel being dissolved in the salt, many MSR designs involve fuel replacement and reprocessing on a (semi-)continuous basis. These operations render standard techniques, wherein macroscopic samples are removed from fuel rods for assaying the actinide content, useless. Thus, MSR designs raise the challenge of how the isotopic composition of the dissolved fuel content will be determined as the reactor burn progresses. Actinide and fission fragment inventories for standard light water reactors are determined by assaying the spent fuel rods. In the case of MSRs, the fuel is dissolved into the salt so that standard technique cannot be used. In this sub-section, we examine a possible method of measuring inventories in-line at the reprocessing station, where the fission gases would be released. The main concept is to replace actinides measurements with actinide *inferences* from the fission gases measurements, applicable to MSRs with continuous refueling and reprocessing operations. Fission gases are automatically released in any reprocessing operation and are easily collected for assay (**Figure 8**).

In MSR, the salt is typically treated in-line to remove as many fission poisons as possible One set of fission products that are relatively easy to remove are the noble gases, xenon, and krypton. This constant separation of the Xe and Kr gases makes the on-site reprocessing plant an ideal station to monitor the isotopic composition of these gases. As the burn proceeds:

- the stable Kr and Xe isotope abundances (except for ^{136}Xe, discussed below) simply increase linearly with burnup,
- the unstable Kr and Xe isotopes come to an equilibrium value that is determined by the cumulative fission yield
- the reactor poison ^{135}Xe asymptotes to an equilibrium value determined by the neutron flux and ^{135}Xe capture cross-section.

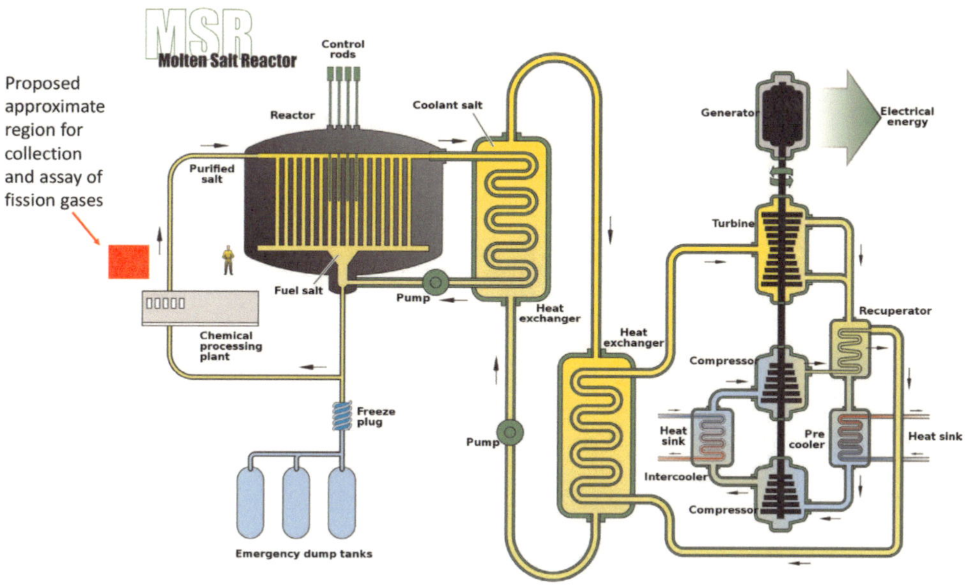

Figure 8.
An example of a molten salt reactor, taken from Wikipedia. The fuel is dissolved in the salt. The red box indicates a proposed online operation where fission gas measurements would be made, that is, next to the reprocessing plant.

Therefore, for example, the growth rate of stable[1] ^{34}Xe is simply,

$$\frac{dN_{134Xe}}{dt} = \Gamma_f \bar{f}_{134Xe}, \tag{4}$$

where Γ_f is the fission rate and $\bar{f}_A$ is the burn-weighted cumulative fission yield of nucleus A. **Figure 9** shows the simple linear increase in the ^{134}Xe/^{135}Xe ratio with the burnup for an MSR with thorium chloride fuel enriched with 10% ^{233}U [37]. Thus, a measurement of the ^{134}Xe/^{135}Xe ratio can be inverted to determine the burnup of the fuel, and when coupled with a reactor simulation, the isotopic content.

4.2 Deducing the grade of plutonium from volatile fission fragments

The grade of plutonium is normally defined to be the ratio ^{240}Pu/^{239}Pu, and weapons-grade and reactor-grade fuel are distinguished with the former typically having a ^{240}Pu/^{239}Pu ratio of less than 7% and the latter a ratio of about 25%. If the reactor is running under equilibrium conditions, the ^{240}Pu/^{239}Pu ratio is only dependent on the total neutron fluence, **Figures 10** and **11**. There is no dependence on the enrichment of the uranium fuel. However, the relationship between fluency and the ^{240}Pu/^{239}Pu ratio has a small flux dependence, especially at high flux. This is because ^{239}Pu is produced by the decay of ^{239}Np, which has a half-life of 2.355 days, and equilibrates at a rate that is flux-dependent, **Figure 2**.

Deducing the neutron fluence (n/cm^2) that fuel has been exposed to, in order to deduce the grade of plutonium, can be translated into a problem of deducing the neutron flux (n/cm^2/sec) and the total irradiation time.

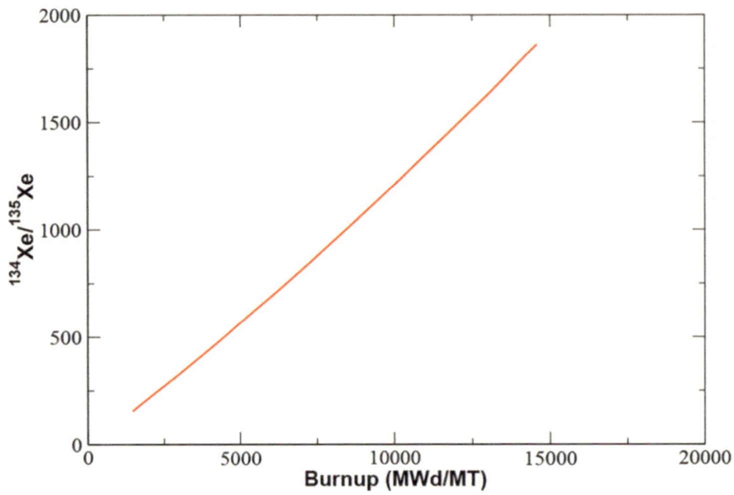

Figure 9.
The increase in the ratio of $^{134}Xe/^{135}Xe$ with burnup for thorium chloride+10% ^{233}U MSR. By measuring the ratio of the stable xenon isotopes to ^{135}Xe, the fuel burnup, and hence the fuel isotopic content can be deduced. [from Monteburns simulations [1, 37].

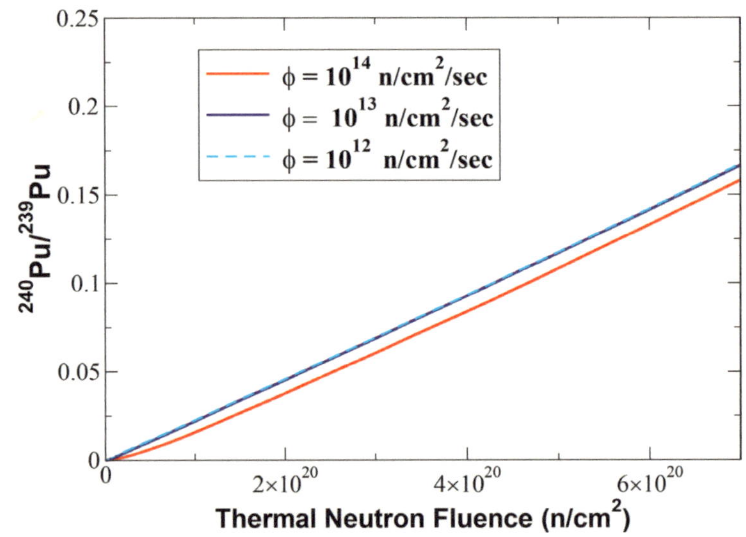

Figure 10.
The $^{240/239}Pu$ ratio dependence on total neutron fluence. For high flux, the curve becomes flux dependent because of the time required for ^{239}Np to reach equilibrium.

4.2.1 Deducing the neutron flux from Xe and Cs fission fragments

The competition between the 9.14-hour decay of xenon-135 to cesium-135 and the thermal neutron capture on xenon-135 to xenon-136, with an abnormally large cross-section of $2.6x10^6$ barns, causes the concentration of the fission products ^{136}Xe and ^{135}Cs to be sensitive functions of the neutron flux, **Figure 12**.

If the thermal neutron flux is low, the ^{135}Xe dominantly beta-decays to ^{135}Cs, while if the thermal flux is high, the neutron capture on ^{135}Xe proceeds faster than the beta-

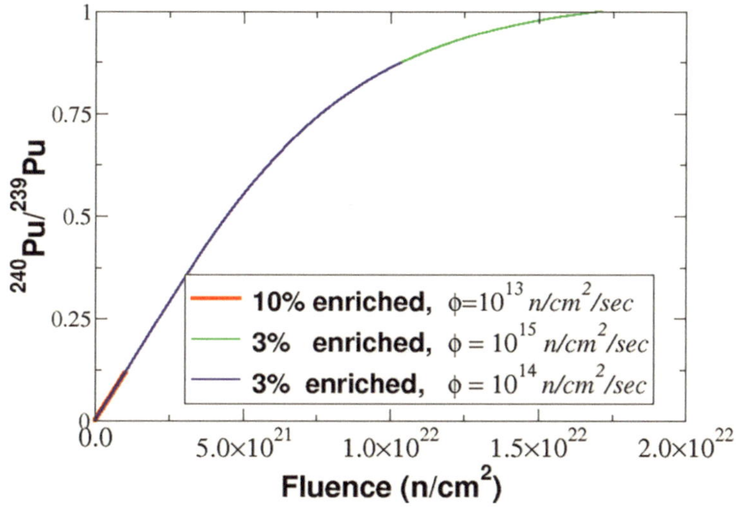

Figure 11.
The $^{240/239}Pu$ ratio is not dependent on the fuel enrichment. In these simulations, the half-life of 239 Np has arbitrarily been set to 1 hour to emphasize the (mostly) simple dependence on fluence.

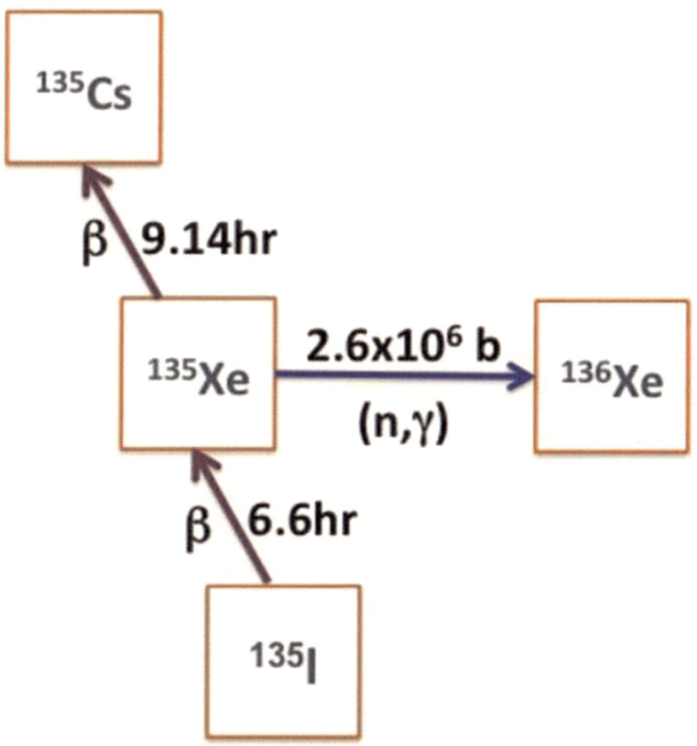

Figure 12.
The concentrations of ^{136}Xe and ^{135}Cs relative to other fission products, such as ^{134}Xe, are determined by the magnitude of the thermal neutron flux. This sensitivity arises because the rate of thermal neutron capture on ^{135}Xe competes with rate of decay of ^{135}Xe to ^{135}Cs.

decay because the capture cross section is of the order of 10^6 barns. In comparison, most fission products are produced directly by fission and/or by the beta-decay of other fission fragments.

^{135}Xe is a well-known reactor poison. Under steady-state conditions, it reaches an equilibrium value that depends on the capture rate of ^{136}Xe and on the concentration of ^{135}I. Changes in the thermal flux cause short-term fluctuations in the level of xenon-135, and after about 40 to 50 hours of steady flux, the xenon-135 level settles to a new equilibrium value that reflects the new flux. Such changes in flux, as well as shut-downs and restarts of the reactor, also affect the ^{136}Xe/^{134}Xe and ^{135}Cs/^{137}Cs ratios.

4.2.2 The dependence of ^{136Xe} on the reactor thermal neutron flux

In a reactor, ^{136}Xe is produced by three main mechanisms: as a direct fission product, from the β-decay of ^{136}I, and from the ^{135}Xe(n,γ)^{136}Xe reaction. The first two mechanisms determine the so-called cumulative fission yield, which is about 7% per fission for both uranium and plutonium.

Under steady-state conditions, the growth rate of 136Xe is [38],

$$\begin{aligned} \dot{N}_{136} &= \bar{f}_{136}\Gamma_f + N_{135Xe}^{equil}\phi\sigma \\ \dot{N}_{136} &= \bar{f}_{136}\Gamma_f\left[1+\frac{\bar{f}_{135Xe}}{\bar{f}_{136Xe}}\frac{\phi\sigma}{\lambda_{135I}+\phi\sigma}\right] \end{aligned} \tag{5}$$

where Γ_f is the fission rate, f_A is the burn-weighted cumulative fission yield of nucleus A, ϕ is the thermal neutron flux, σ the capture cross section on ^{135}Xe, and N^{equil} represents the ^{135}Xe equilibrium value.

The ^{136}Xe production rate falls between two limits,

$$\begin{aligned} \phi\sigma \gg \lambda_{135I} &: \dot{N}_{136} = \bar{f}_{136}\Gamma_f\left(1+\frac{\bar{f}_{135Xe}}{\bar{f}_{136Xe}}\right) \\ \phi\sigma \gg \lambda_{135I} &: \dot{N}_{136} = \bar{f}_{136}\Gamma_f \end{aligned} \tag{6}$$

The ratio of xenon-136 to xenon-134 is then,

$$\frac{N_{136}}{N_{134}} = \frac{\bar{f}_{136Xe}}{\bar{f}_{134Xe}}\left(1+\left(\frac{\bar{f}_{135Xe}}{\bar{f}_{134Xe}}\right)\frac{\phi\sigma}{\lambda_{135}+\phi\sigma}\right) \tag{7}$$

From Eq. 7, the ^{136}Xe/^{134}Xe ratio equilibrates on a (flux-dependent) time scale of weeks. **Figure 6** displays the situation for different values of the thermal flux (**Figure 13**).

4.2.3 The relationship between thermal flux and the ^{135}Cs/^{137}Cs ratio

In this section, we derive an analytic expression relating the concentration of cesium-135 to the thermal neutron flux. Cesium-135 is produced through the decay chain 135I ➔ 135Xe ➔ 135Cs, which is complicated by the transmutation of xenon-135 *via* neutron capture. The number of cesium-135 atoms produced in a time t is given by,

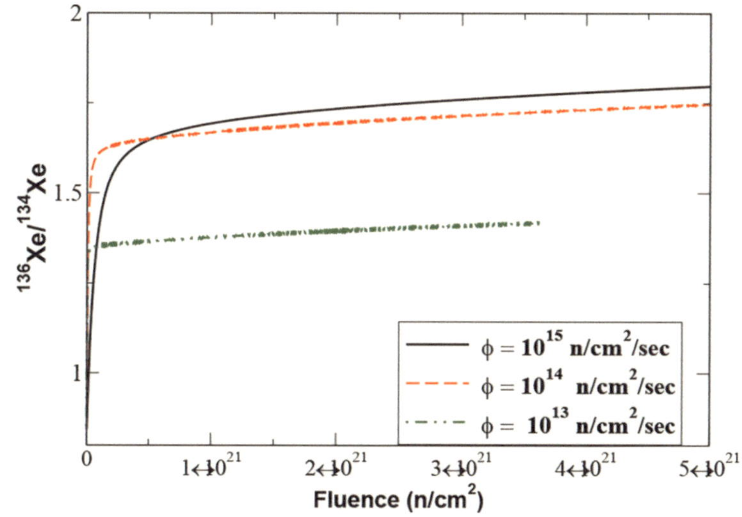

Figure 13.
The $^{136}Xe/^{134}Xe$ ratio equilibrates to a value that is determined by the thermal neutron flux. Thus, this ratio can be used to deduce the flux. The residual slope of the plateaus in the curves is caused by the slow evolution of the fuel composition, and hence the cumulative fission yields, from ref. [38].

$$N_{135Cs}(t) = \lambda_{135Xe} \int_0^t dt' \left[\frac{\frac{\bar{f}_{135I}}{\bar{f}_{137Cs}} \dot{N}_{137Cs} - \dot{N}_{135I} - \dot{N}_{135Xe}}{\lambda_{135Xe} + \phi\sigma} \right] \tag{8}$$

Here $\dot{N}_A$ is the growth rate of nuclide A.

An analytic expression for $^{135}Cs/^{137}Cs$ ratio can be obtained by assuming that the concentration of iodine-135 and xenon-135 are at their equilibrium values and that the thermal neutron flux is constant. Under these assumptions,

$$N_{135Cs} = \frac{\lambda_{135Xe}}{\lambda_{135Xe} + \phi\sigma} \frac{\bar{f}_{135I}}{\bar{f}_{137Cs}} N_{137Cs} + \frac{\phi\sigma}{\lambda_{135Xe} + \phi\sigma} \left(N_{135I}^{equil} + N_{135IXe}^{equil} \right)$$

$$N_{135Cs} = N_{137Cs} \frac{\bar{f}_{135I}}{\bar{f}_{137Cs}} \lfloor \frac{\lambda_{135Xe}}{\lambda_{135Xe} + \phi\sigma} + \frac{\phi\sigma}{\lambda_{135Xe} + \phi\sigma} \left(\frac{1}{\lambda_{135I} T_{irrad}} \right) \left(1 + \frac{\lambda_{135I}}{\lambda_{135Xe} + \phi\sigma} \right) \rfloor \tag{9}$$

Here λ_{135I} is the decay constant for iodine-135, and T_{irrad} is the total irradiation time in the reactor. The form of the second term, which gives rise to the term inversely proportional to the total irradiation time, arises from the proper treatment of startup transients.

4.2.4 The effect of reactor shutdowns

After a reactor shutdown, the operators must wait long enough to ensure that all the ^{135}Xe and ^{135}I have decayed before restarting the reactor. Failure to do so was one of the errors that occurred in the Chernobyl accident. Each time the reactor is shut down, the concentration of ^{135}Cs is increased. For constant neutron flux operation, shutdown adds P terms identical to the second term in Eq. 9, where P is the number of shutdowns before the fuel is removed. The minimum value of P is one, and the xenon ratio becomes,

$$\frac{N_{136Xe}}{N_{134Xe}} = \frac{\bar{f}_{136Xe}}{\bar{f}_{134Xe}} \lfloor 1 + \frac{\phi\sigma}{\lambda_{135Xe} + \phi\sigma} \left(\frac{\bar{f}_{135Xe}}{\bar{f}_{136Xe}}\right) \left(1 - \frac{P}{\lambda_{135I} T_{irrad}^{total}}\right) \left(1 + \frac{\lambda_{135I}}{\lambda_{135Xe} + \phi\sigma}\right) \rfloor, \tag{10}$$

while the cesium ratio becomes,

$$N_{135Cs} = N_{137Cs} \frac{\bar{f}_{135I}}{\bar{f}_{137Cs}} \lfloor \frac{\lambda_{135Xe}}{\lambda_{135Xe} + \phi\sigma} + \frac{\phi\sigma}{\lambda_{135Xe} + \phi\sigma} \left(\frac{\mathrm{P}}{\lambda_{135I} T_{irrad}^{total}}\right) \left(1 + \frac{\lambda_{135I}}{\lambda_{135Xe} + \phi\sigma}\right) \rfloor. \tag{11}$$

Here T_{irrad}^{total} is the sum of all the irradiation times to which the fuel was exposed. The shutdown correction for the cesium ratio, but not the xenon ratio, becomes significant at high flux. The difference is that for xenon the correction is relative to the constant 1. However, a measurement of the cesium ratio in spent fuel contains information on both the flux and the number of shutdowns.

4.2.5 Comparisons between the theoretical and experimental cesium and xenon ratios

Maeck *et al.* measured [39] fission isotope product ratios by irradiating highly enriched uranium targets in the Advanced Test Reactor (ATR) and in the Engineering Test Reactor (ETR) at Idaho National Laboratory. By placing the targets at different distances from the reactor mid-plane, each target was exposed to a different flux, and the flux ranged between 6×10^{12} and 1.5×10^{14} n/cm^2/sec. The ATR targets were 93% ^{235}U and were irradiated for 100 days over 335 days, with several shutdowns. The ETR targets were more than 99% pure ^{235}U and one set of targets was irradiated for 20 days and another for 180 days over 320 days with some unspecified number of shutdowns.

Figures 14 and **15** shows the experimental data and the calculated ^{135}Cs/^{137}Cs ratios. As can be seen, the ^{137}Cs/^{135}Cs ratio scales approximately linearly with the flux, but the reactor shutdowns cause the isotopic ratio to increase with flux at a slower rate at high fluxes. **Figure 15** shows the flux dependence of the ^{136}Xe/^{134}Xe ratio, which is a sensitive diagnostic at low, but not high, fluxes. The ^{136}Xe/^{134}Xe ratio is an ideal probe for low-flux graphite reactors, for example.

4.2.6 Monitoring reactor on-off times

To use the Cs and Xe ratios to address our main problem, namely, the plutonium grade of fuel that is being reprocessed, there remains the problem of knowing the irradiation times and the number of shutdowns. There are several techniques and the usefulness of these depends on the standoff distance from the reactor. We discuss only one method here, namely, reactor neutrino monitoring.

Several neutrino experiments have shown very successfully that measurements of the total number of neutrinos emitted from a reactor can be used to monitor reactor on-off times. Two examples are shown in **Figure 16**.

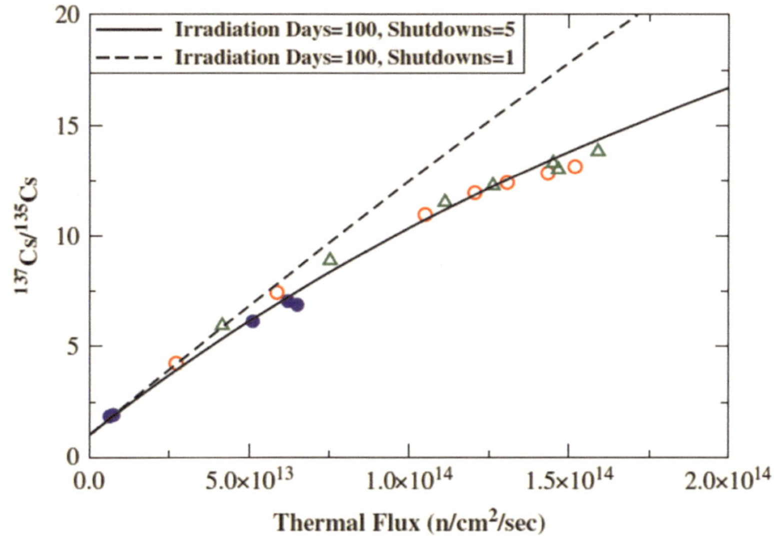

Figure 14.
The $^{137}Cs/^{135}Cs$ ratio. The measurements are the data of Maeck et al. [39]. The deviation of the data from the one-shutdown scenario shows the strong dependence of this ratio on the number of shutdowns. The different colors used to display the data points distinguish the different reactor or irritation times used. The open circles and triangles correspond to targets irradiated at the ATR reactor, while the blue circles are data from targets in the ETR reactor.

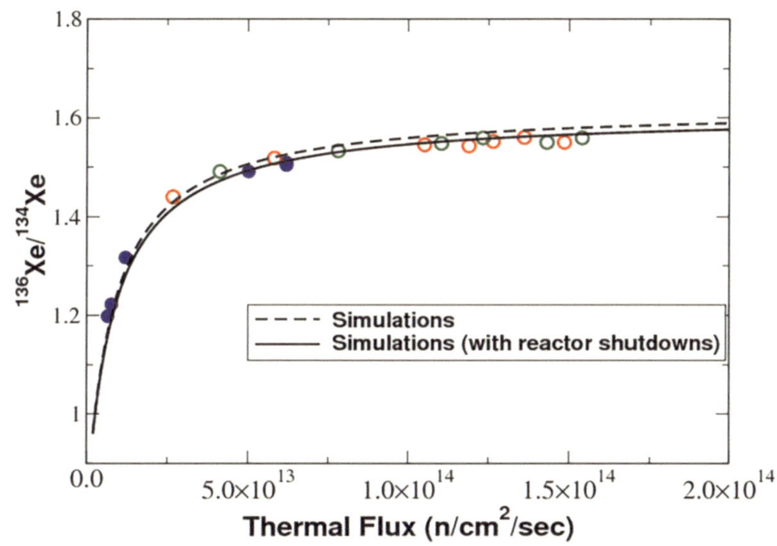

Figure 15.
The $^{136}Xe/^{134}Xe$ ratio. The measurements are the data of Maeck [39]. Reactor shutdowns have very little effect on this ratio. The xenon ratio is a very good diagnostic for low-flux, but not for high-flux, reactors.

Thus, the grade of reactor fuel can be determined if:

a. The original unirradiated fuel is some form of uranium 238 U + 235 U but of unknown enrichment.

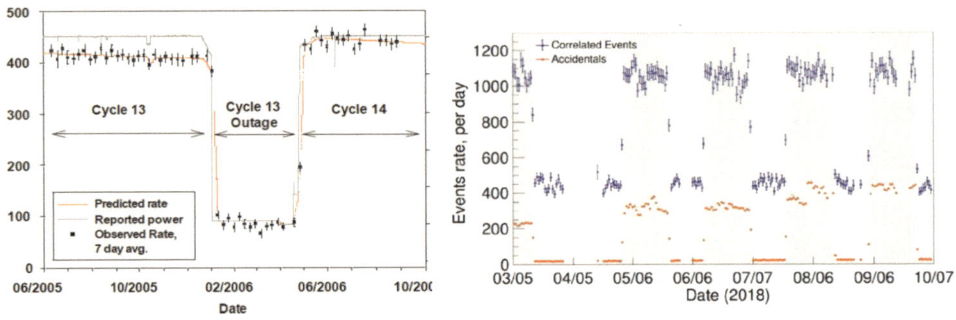

Figure 16.
Antineutrino monitoring of reactors allows accurate measurements of the reactor on and off burn times. (left) form the SONGS detector data taken at the San Onofre reactor [40]. (right) from the PROSPECT detector data taken at the 85 MW HFIR reactor at oak ridge [41].

b. During the entire burn of the fuel antineutrino, monitoring was sufficiently accurate to determine when the reactor was on or off.

c. Fission gases produced in the fuel during the burn can be measured.

This is a nice example of combining two diagnostics to gain enhanced information for nonproliferation, namely, combining reactor fission gas data and antineutrino data.

5. Fission applications to nuclear medicine

This section is intended as a short survey of the most common nuclear medical procedures that were derived from fission. In general, uses of radioisotopes and nuclear techniques in medicine fall into two main classes: (a) imaging and diagnosing structures in the body and their functioning, and (b) treating diseases. Fission isotopes contribute to both classes and many reactor facilities around the world are involved in reprocessing spent fuel to produce medical isotopes. The advances in nuclear medicine involve highly multi-disciplinary research and biology and physiology are an important aspect of these studies because of the importance of accurately targeting the organ of interest.

5.1 Nuclear imaging

The ^{99}Mo-^{99m}Tm pair are the most used medical isotope and ^{99m}Tc is used for imaging and it can be delivered to organs or tissues when injected into the body. Nuclear imaging can detect biochemical changes in an organ (as opposed solely to changes in size), and, thus, can be used to affect decisions in treating disease. The imaging technique used is known as single photon emission computed tomography (SPECT). SPECT [42] uses gamma rays emitted in the decay of a radionuclide. The ^{9m}Tc used in SPECT comes from the 2.75-day decay of the fission product ^{99}Mo. Most of the ^{99}M is produced in reactors that use highly enriched uranium (HEU) targets, although there is considerable research on moving to production using low-enriched uranium. The isomer of ^{99}Tc has a half-life of 6 hours and emits a 140 keV photon. Some fractions of the photons reach the detector after escaping the body and can be used to create a 3-D image. This is done by taking 2-D projections at different angles. The 3-D dataset can be manipulated into slices along any axis. One of the advantages of SPECT is that it can be used to study blood flow. The temporal resolution for SPECT is limited so that only averaged views in time are possible. There has been some discussion in the literature [43, 44] about the advantages or disadvantages of SPECT versus positron emission tomography (PET). However, SPECT myocardial perfusion imaging is used almost 20 times more frequently than PET in clinical practices in the United States [44].

5.2 Radionuclide therapy

In targeted radionuclide therapy (TRT) [**45**], radiopharmaceuticals are delivered to tumor cells by targeting specific characteristics of the tumor, such as antigens. In this way, a toxic level of the radionuclide can be delivered to the site. The radionuclide of choice emits charged particles with the required stopping range for effective destruction of malignant tissue, and includes, beta electrons, Auger electrons, and alpha particles. A commonly used fission isotope is ^{90}Y. As a high-energy beta-emitter, ^{90}Y is used for the treatment of larger tumors and is a successfully targeted radiotherapy of the liver. Another application of TRT is in the treatment of bone cancers. A large fraction of breast and prostate cancer patients develop bone metastases, which can cause severe pain. Both ^{153}Sm and ^{89}Sr deliver high radiation doses to bone metastases and micrometastases in the bone marrow. The field of TRT is ever going, with fission fragments playing major role. For example, ^{131}I, which has traditionally been used to treat thyroid cancer, is now being studied for TRT in metastatic melanoma [**46**] (**Figure 17**).

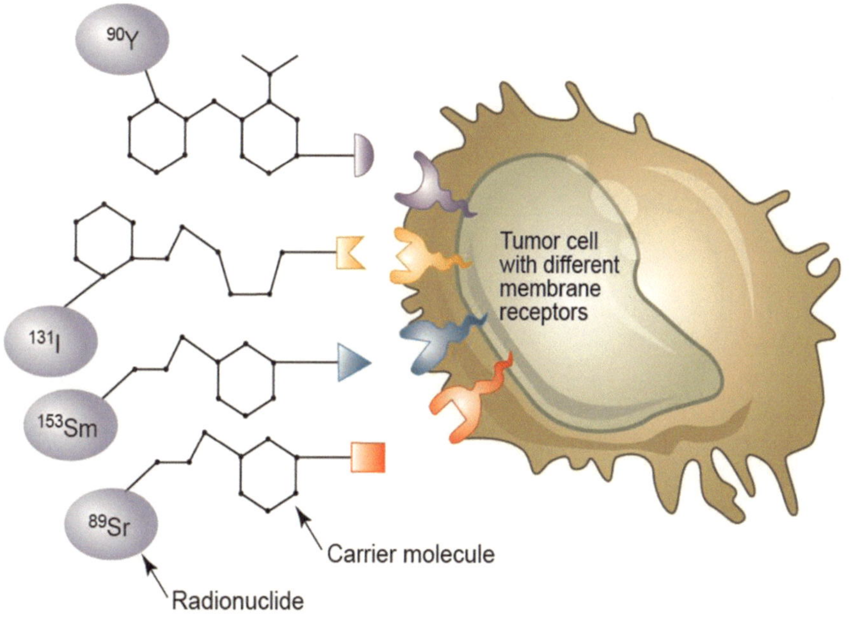

Figure 17.
A toxic level of radiation is delivered to a diseased site by attaching a radionuclide to a molecular carrier that binds to the site or tumor.

6. Concluding remarks

Nuclear fission has and continues to be applied to a very broad range of needs. Many of these applications have developed into entire subfields of their own and have their own dedicated technical journals. This chapter includes a discussion of nuclear reactors, fission neutrino oscillation studies, nuclear nonproliferation, and nuclear medicine. However, there are many more applications of nuclear fission, some of which are discussed in Ref. [47]. Nuclear fission is very important for other fields, such as nuclear astrophysics, geophysics, nuclear submarines, and nuclear propulsion through space. No doubt the fission of applied nuclear fission will continue to grow as a healthy scientific area.

Author details

Anna C. Hayes
Theoretical Division, Los Alamos National Laboratory, Los Alamos, NM, USA

*Address all correspondence to: anna_hayes@lanl.gov

References

[1] Trellue H. Los Alamos national laboratory internal report, development of Monteburns: A code that links MCNP and ORIGEN2 in an automated fashion for burnup calculations. Thesis. 1998; **LA-13514-T**:1-190

[2] Nieto MM, Hayes AC, Wilson WB, Teeter CM, Stanbro WD. Detection of antineutrinos for nonproliferation. Nuclear Science and Engineering. 2005; **149**(3):270-276. DOI: 10.13182/NSE05-A2493

[3] Wilson WB, England TR, Ozer O, Wessol DE. Actinide decay power. Transactions of the American Nuclear Society. 1979;**32**:737

[4] Wilson WB, England TR, LaBAUVE RJ, Mitchell JA. Calculated radionuclide inventories of high-exposure LWR fuels. Nuclear Safety. 1988;**29**:177

[5] Torgerson DF, Shalaby BA, Pang S. CANDU technology for generation III+ and IV reactors. Nuclear Engineering and Design. 2006;**236**:1565-1572

[6] Thorium fuel cycle- Potential benefits and challenges. IAEA-TECDOC-1450. 2005

[7] Status and Advances in MOX Fuel Technology. Technical Reports Series 415. Vienna: IAEA; 2003

[8] Hayes AC, Trellue HR, Nieto MM, Wilson WB. Antineutrino monitoring of burning mixed oxide plutonium fuels. Physical Review C. 2012;**85**:024617

[9] Kulesza JA, Adams TR, Armstrong JC, Bolding SR, Brown FB, Bull JS, et al. MCNP® Code Version 6.3.0 Theory & User Manual. LA-UR-22-30006, Rev. 1. Los Alamos, NM, USA: Los Alamos National Laboratory Tech. Rep; 2022

[10] Wilson WB, Cowell ST, England TR, Hayes AC, Moller P. A Manual for CINDER'90 Version 07.4 Codes and Data. Los Alamos, NM: Los Alamos National Laboratory, LA-UR-07-8412; 2007

[11] Reines F, Cowan CL Jr. Detection of the free neutrino. Physics Review. 1953; **92**:830

[12] Reines F, Cowan CL Jr. Free antineutrino absorption cross section. I. Measurement of the free antineutrino absorption cross section by protons. Physica Review. 1959;**113**:273

[13] The Reines-Cowan Experiments. Detecting the poltergeist. Los Alamos Science. 1997;**25**:3

[14] Pontecorvo B. Neutrino experiments and the problem of conservation of Leptonic charge. Zhurnal Eksperimental'noi i Teoreticheskoi Fiziki. 1968;**53**:1717

[15] Mention G, Fechner M, Lasserre T, Mueller TA, Lhuillier D, Crobier M, et al. The reactor antineutrino anomaly. Physics Review. 2001;**D83**:073006

[16] Apollonio M et al. Limits on neutrino oscillations from the CHOOZ experiment. Physics Letters B. 1999;**466**:415

[17] Boehm F et al. Search for neutrino oscillations at the Palo Verde nuclear reactors. Physical Review Letters. 2000; **84**:3764

[18] Boehm F et al. Results from the Palo Verde neutrino oscillation experiment. Physical Review D. 2000;**62**:072002

[19] *Eguchi K* et al. (*KamLAND Collaboration*). "First results from

Kam LAND: Evidence for reactor antineutrino disappearance. Physical Review Letters. 2003;**90**(2): 021802–021807

[20] Daya Bay Collaboration. Observation of electron-antineutrino disappearance at Daya bay. Physics Review. 2012;**108**(17):171803

[21] RENO Collaboration. Observation of electron-antineutrino disappearance at RENO. Physical Review Letters. 2012; **108**(18):191802

[22] Abe Y et al. (Double Chooz Collaboration) "Improved measurements of the neutrino mixing angle θ_{13} with the double Chooz detector". Journal of High Energy Physics. 2014;**86**(10)

[23] Huber P. Determination of antineutrino spectra from nuclear reactors. Physical Review C. 2011;**84**: 024617

[24] Huber P. Erratum. Physical Review C. 2012;**85**:029901

[25] Mueller TA et al. Improved predictions of reactor antineutrino spectra. Physical Review C. 2011;**83**: 054615

[26] An FP et al. Evolution of the reactor antineutrino flux and spectrum at Daya Bay. Physical Review Letters. 2017;**118**: 251801

[27] Hayes AC, Vogel P. Reactor Neutrino Spectra. Annual Review of Nuclear and Particle Science. 2016;**66**: 219-244

[28] Schopper HF. Weak Interactions and Nuclear Beta Decay. Amsterdam: North-Holland; 1966

[29] Schreckenbach K, Faust HR, von Feilitzsch F, Hahn AA, Hawerkamp K, Vuilleumier JL. Absolute measurement of the beta spectrum from 235-U fission as a basis for reactor antineutrino experiments. Physics Letters. 1981;**99B**: 251

[30] von Feilitzsch F, Hahn AA, Schreckenbach K. Experimental beta-spectra from ^{239}Pu and ^{235}U thermal neutron fission products and their correlated antineutrino spectra. Physics Letters. 1982;**118B**:162

[31] Schreckenbach K, Colvin G, Gelletly W, von Feilitzsch F. Determination of the antineutrino spectrum from ^{235}U thermal neutron fission products up to 9.5 MeV. Physics Letters. 1985;**160B**:325

[32] Hahn AA, Schreckenbach K, Gelletly W, von Feilitzsch F, Colvin G, Krusche B. Antineutrino spectra from ^{241}Pu and ^{239}Pu thermal neutron fission products. Physics Letters B. 1989; **218**:365

[33] Hayes AC, Gerard Jungman EA, McCutchan AA, Sonzogni GT, Garvey XBW. Analysis of the Daya bay reactor antineutrino flux changes with fuel burnup. Physical Review Letters. 2018;**120**:022503

[34] Kopeikin V, Skorokhvatov M, Titov O. Reevaluating reactor antineutrino spectra with new measurements of the ratio between ^{235}U and ^{239}Pu β spectra. Physical Review D. 2021;**104**:L071301

[35] Kellett MA, Bersillon O, Mills RW. The JEFF-3.1/−3.1.1 radioactive decay data and fission yields sub-libraries. JEFF Report 20, NEA Report No. 6287. 2009:1-147

[36] Chadwick MB et al. The ENDF/B-VIII.0 library. Nuclear Data Sheets. 2011;

112:2887. ISBN 92–0–103405–9. ISSN 1011-4289

[37] Trellue H, Lafreniere P, Maldonado A, Hayes-Sterbentz A, Mehta V. Calculation of Fission Gases from Different Fuel Sources in Molten Salt Reactors, LANL internal report, unpublished

[38] Hayes AC. Gerard Jungman, determining reactor flux from xenon-136 and cesium-135 inn spent fuel. Nuclear Instruments and Methods in Physics Research A. 2012;**690**:68

[39] Maeck WJ, Tromp LR, Duce FA, Emel WA. Isotope Correlation Studies Relative to High Enrichment Test Reactor Fuels, ICP-1156. Idaho: Allied Chemical Corporation; 1978

[40] Bowden N et al. Experimental results from an antineutrino detector for cooperative monitoring of nuclear reactors. Nuclear Instruments and Methods A. 2007;**572**:985-998

[41] Andriamirado et al. Improved short-baseline neutrino oscillation search and energy spectrum measurement with the PROSPECT experiment at HFIR. PRD103. 2021;**03201**:03201-1-03201-45

[42] Seo Y, Mari C, Hasegawa BH. Technological development and advances in single-photon emission computed tomography/computed tomography. Seminars in Nuclear Medicine. 2008;**38**:177-198

[43] Shrestha U, Sciammarella M, Alhassen F, Yeghiazarians Y, Ellin J, Verdin E, et al. Measurement of absolute myocardial blood flow in humans using dynamic cardiac SPECT and 99mtc-tetrofosmin: Method and validation. Journal of Nuclear Cardiology. 2017;**24** (1):268-277. DOI: 10.1007/s12350-015-0320-3

[44] Slomka P, Berman DS, Germano G. Myocardial blood flow from SPECT. Journal of Nuclear Cardiology. 2017;**24** (1):278-281

[45] Williams LE, DeNardo GL, Meredith RF. Targeted radionuclide therapy. Medical Physics. 2008;**35**: 3062-3068

[46] Thivat E et al. Phase I study of [^{131}I] ICF01012, a targeted radionuclide therapy, in metastatic melanoma: MELRIV-1 protocol. BMC Cancer Actions. 2022;**22**:1. DOI: 10.1186/s12885-022-09495-3

[47] Hayes AC. Applications of nuclear physics. Reports on Progress in Physics. 2017;**80**:026301

Chapter 8

Basics of Neutron Imaging

Eberhard H. Lehmann

Abstract

Neutron imaging is established at many neutron sources around the world as a method for noninvasive investigations of samples and object on the macroscopic scale. Similarly to X-ray imaging, it provides the possibility to "look through" materials and allows one to "see" the inner, hidden content. However, owing to the complete different interaction mechanism, neutron imaging provides very different and complementary contrasts compared to X-rays, even if the image quality often is about the same. We report about the method's principles, describe the state of the art, and give an outlook for new trends and developments.

Keywords: neutron source, research reactor, neutron tomography, nondestructive testing, neutron detection, neutron moderation, thermal neutrons, cold neutrons

1. Introduction

The discovery of neutron in 1932 and afterward the process to initiate nuclear fission in a chain reaction provided the opportunity to set free an enormous amount of energy for two purposes: nuclear weapons and nuclear power plants. However, there is a third aspect in using fission reactions: the generation of fields of free neutrons.

For this purpose, the erection of specialized strong neutron sources, the so-called research reactors, which have been designed and built in many countries all over the world, has started just after World War II. Their principle is the fission of U-235 in a compact nuclear core with the purpose of highest possible neutron output, but minimal energy production. In this case, the energy release is seen to be more disturbing than useful. In the end, the necessity to remove the resulting energy from the reactor core limits the strength of such neutron sources to about 100 MW and the corresponding neutron flux level of about 10^{15} cm^{-2} s^{-1}.

Such strong neutron sources have been used for many scientific and also practical purposes until today. Next to the training of specialists in the reactor technology for power plants and the test of nuclear fuel in the compact cores with its high power density, the neutrons themselves are the target and tool of research.

In the interaction of neutrons with matter, nuclear reactions are initiated with the result of new isotopes, often not existing in nature due to their limited lifetime during decay. Many of such isotopes can be used in medicine for diagnostics and therapy and as gamma sources for several practical applications. On the other hand, neutrons are scattered at atomic nuclei and provide in this way important information about the investigated materials, not accessible by other scientific methods.

Neutron imaging, the topic of this chapter, is a method on the macroscopic scale, performed with the aim to study materials, too. Similarly to common X-ray diagnostics, the material under investigation is illuminated by a directed beam of neutrons, and the detector behind registers the remaining neutron intensity. By comparison to the beam intensity without sample, the amount of missing (captured, scattered out) neutrons can be determined and related to the attenuating properties of the investigated material.

In order to perform neutron imaging experiments, next to the neutron source, a dedicated and well-designed facility has to be built. Such neutron imaging stations are part of the suite of experimental devices around the neutron source (up to now mainly research reactors).

In this chapter, we will describe how the best conditions for neutron imaging are provided: from the initial fission reaction, the tuning of the suitable neutron energy in the moderation process, via the beam extraction toward the best detector design. We will show some of the best installations for neutron imaging, describe the principles of the interaction of neutrons with matter, and demonstrate with examples how neutron imaging can be used for scientific and technical purposes. In the outlook, we will sketch how neutron imaging is developed further by methodical improvements and new neutron sources next to the still existing research reactors.

2. Reactor-based neutron sources for neutron imaging

As described above, none of the research reactors have been built exclusively for neutron imaging. Other applications like isotope production, neutron scattering, or fuel development have higher priorities and importance. Therefore, not all neutron sources are able to perform neutron imaging on an up-to-date level yet. The main aspect is the access to a suitable beam port, where the most useful neutrons can be extracted and a beam is formed for the imaging process.

While the neutrons, resulting from the fission reaction, are very fast (energies on the order of MeV), the most useful and interesting neutrons are very slow (in the range of meV). The reason to use slow neutrons is their specific interaction properties with matter (neutron's wavelength are about the atomic distances in solids), high detection probability, and high diversity in the cross-sectional data, which allows one to distinguish even individual isotopes very precisely.

In order to slow down fission neutrons over such a high energy range (nine orders of magnitude), a moderation process is necessary. It has been found out that the elastic collision with very light materials, which do not absorb neutrons much, enables a stepwise reduction of the neutron energy. The lower the mass of the moderator material, the less nuclear collisions are needed to reach the intended "thermal energy"—the equilibrium with the thermal movements of the moderator's molecules. Hydrogen is the ideal material to provide such moderation performance. However, the natural hydrogen ^{1}H also absorbs neutrons. This reduces the amount of the resulting slow neutrons. Much better is the hydrogen isotope deuterium ^{2}H = the nucleus with a proton and an additional neutron. It absorbs nearly no neutrons, but needs some more collisions to arrive at the required thermal equilibrium.

About 2.4 fast neutrons are emitted with each fission reaction of U-235, of which already one neutron is needed to continue the chain reaction. From the remaining 1.4 neutrons, a part is lost by parasitic capture in the moderator or in structural materials in the reactor core. However, there are still enough neutrons remaining, forming a cloud around the reactor core. Because neutrons are neutral particles without a charge, there are no methods available to guide them with electromagnetic fields into a desired direction. Therefore, neutrons can only be extracted from the source region by a selection of useful ones, which fly into the dedicated direction of an experimental facility. Neutron flux intensities of only about 10^7 cm^{-2} s^{-1} are common at imaging stations.

Despite the success of research reactors, spallation neutron sources have been built as replacements and complements, useful in particular with their pulsing option for time-of-flight (TOF) investigations.

3. Principles and options in neutron imaging

Indeed, right with the availability of neutron sources, several attempts were made to use neutrons for imaging purposes. It was quickly found out that such transmission images look similar to X-ray and gamma images, however with alternative contrast features [1, 2].

As sketched in **Figure 1**, neutron images are obtained employing a two-dimensional area detector behind an object when a collimated neutron beam is sent through it. In the beginning of such investigations, photographic film methods were applied. However, as neutrons cannot directly activate the film, a neutron-to-radiation converter has to be used. There are some materials with high absorption probability as Gd, Dy, In, or Au, converting the captured neutrons into another kind of radiation, which then excites the film.

The film techniques were used until the 1990s of the last century (and even today for special certified applications) before some "revolution" happened: the introduction of digital neutron imaging detection systems. In this way, the previous limitations with film were overcome, and new techniques in neutron imaging have been developed and established. **Figure 2** provides an overview on which kind of methods are available today, all based on the advantages of the digital imaging option.

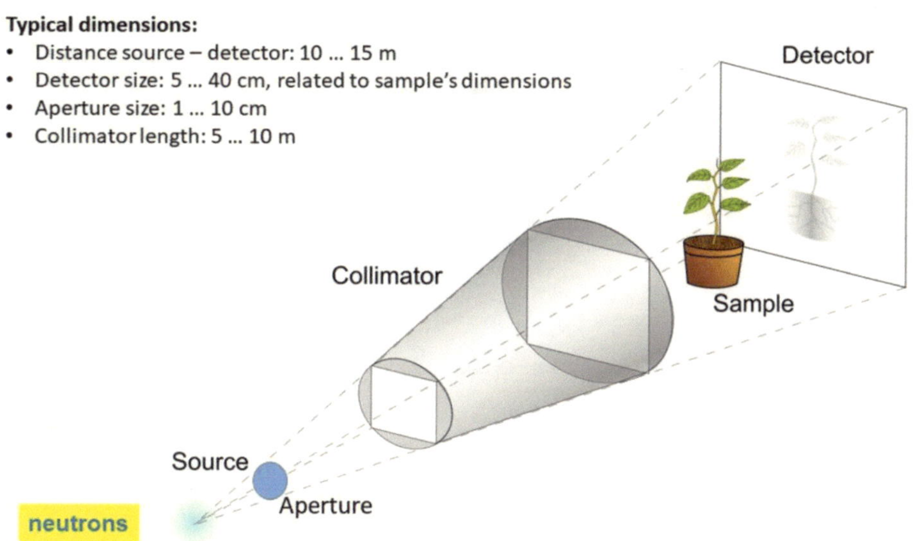

Figure 1.
Sketch of a neutron imaging experiment (not to scale), where an aperture and the collimator define the beam properties.

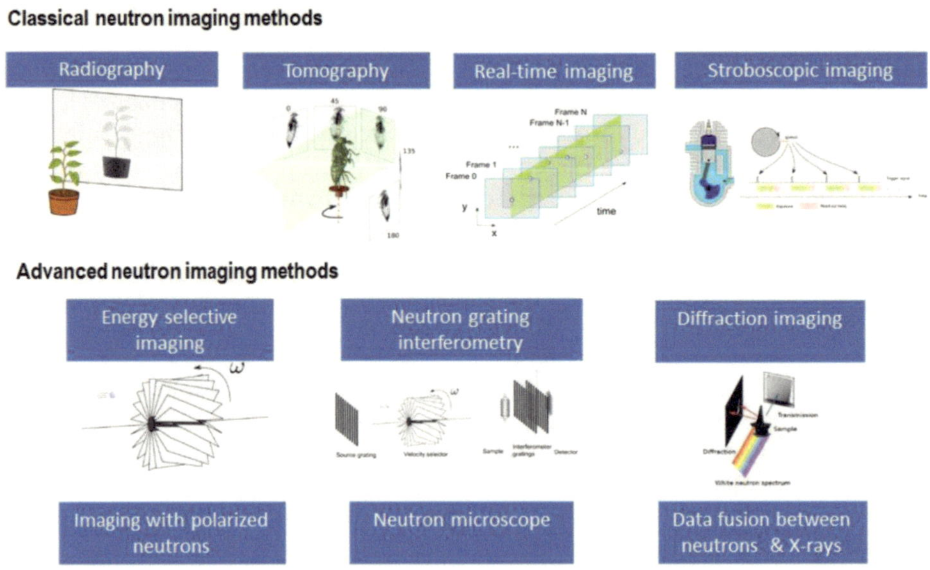

Figure 2.
Overview of neutron imaging techniques, available in several user facilities (see the list in the appendix).

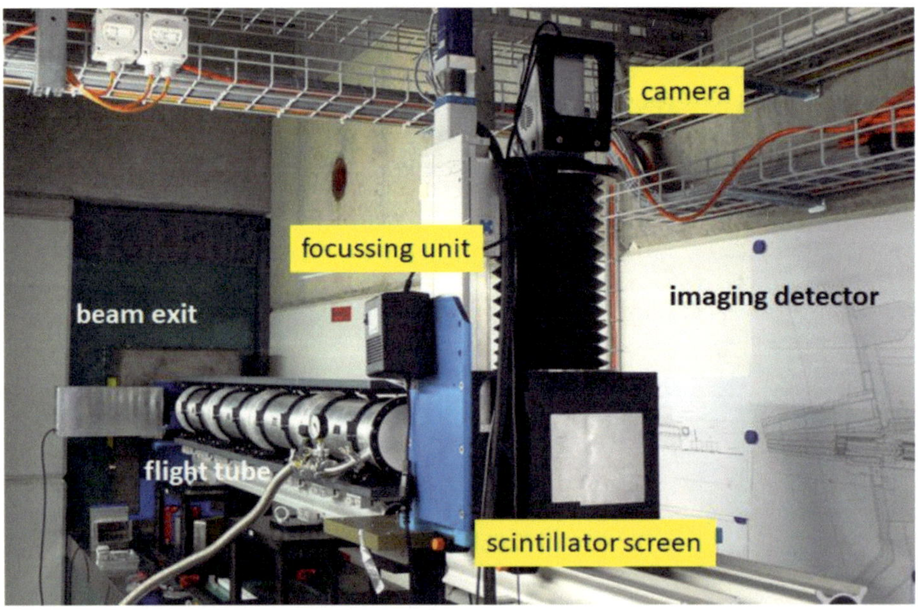

Figure 3.
Setup of a camera-based neutron detector with its major components; the induced light emission from the neutron-sensitive scintillator screen is sent to the camera via a mirror arranged in 45° inside a light-tight box with permission by the IAEA [3].

In particular, the acquisition time for a "valid" neutron image is now on the order of seconds and below, enabling the sampling high amounts of data—to be treated accordingly and stored efficiently.

From the various digital methods, the setup with a highly sensitive digital camera has been found most flexible and efficient. The principle construction is shown in **Figure 3**. Next to the (often expensive) high-performance camera (CCD or CMOS), the neutron-sensitive scintillator screen is of importance. It converts the capture of neutrons into visible light, registered pixelwise in the camera detector. It depends on the particular application if high spatial resolution, high frame rate, or highest quantitative accuracy has to be obtained in the experiments. Accordingly, the composition, layer thickness, and converter material of the scintillator have to be chosen well.

Based on the digital imaging data, neutron tomography is established as a routine method, providing access to the 3D volume of an object, as shown in the example in **Figure 4**. In time sequences, a current image can be compared to a previous one, while the precision to derive values of changes is much enhanced. This is shown in

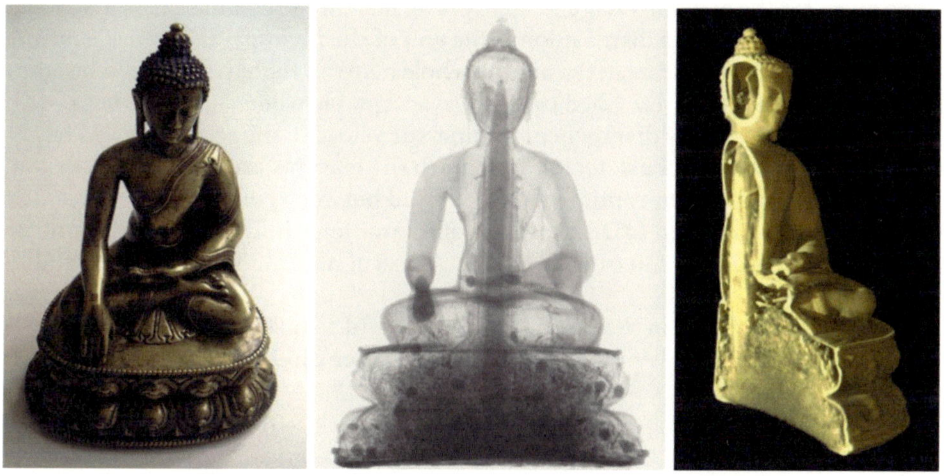

Figure 4.
Results of radiography (middle) and tomography (right) measurements with thermal neutrons, when a bronze sculpture from Tibet (seventeenth century) has been investigated concerning its hidden organic content (wood, paper, and dry flowers).

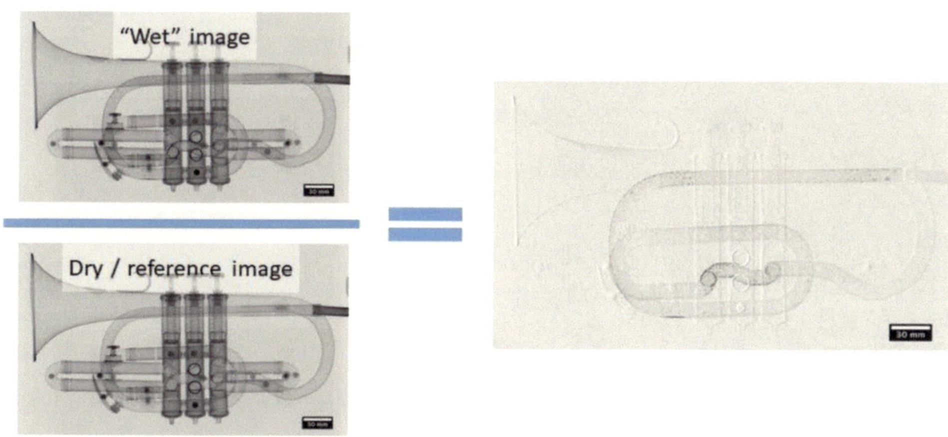

Figure 5.
Using a pixelwise referencing image procedure, it is possible to visualize the moisture accumulation inside the trumpet (online during "playing").

an example in **Figure 5**, where the moisture accumulation during trumpet play is determined very locally and with high sensitivity for water deposits.

4. Generic setup of a neutron imaging facility

Although no neutron imaging facility is constructed identical to another one, the most important components of it can be specified, as shown in **Figure 6**. Starting with the primary source of fast neutrons, the moderators define the neutron energy and the best point for the neutron extraction. Because a high amount of gamma radiation is emitted during the fission process in addition to the neutrons, a direct view on the fission region should be avoided. It is better to look toward the peak of moderated neutrons next to the nuclear core. Remaining gamma radiation can be reduced in its intensity with suitable filters.

The beam for the neutron imaging facility is formed in the collimators in a way to get a quasi-parallel flat neutron distribution at the end of the flight path. Because neutrons interact also with the moisture in the air, the whole neutron flight path should be empty, which means evacuated. A He-filled tube is also acceptable when vacuum is not possible.

To get a beam with low divergence, the aperture close to the source should be small and the distance to the detector long. Both the features are reducing the beam intensity much, and a compromise has to be found between collimation and resulting neutron flux. The so-called L/D-ratio (L = collimator length, D = aperture diameter) should be on the order of 200 or even higher. A useful neutron flux level is on the order of 10^6 cm^{-2} s^{-1}.

To handle the samples in the beam, it is very useful to install remotely driven manipulators, which enable precise motions in all three directions and additionally

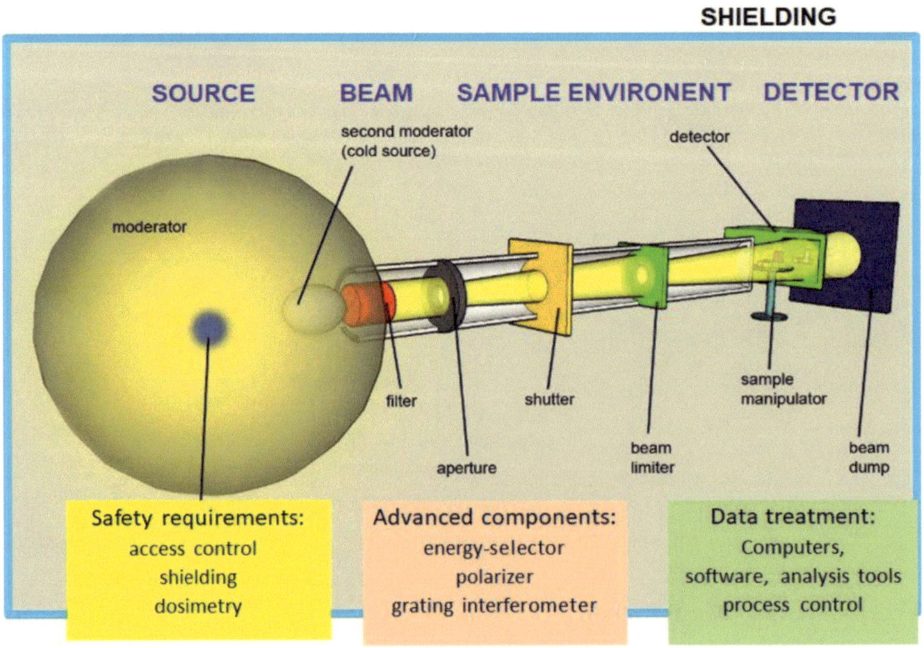

Figure 6.
Generic neutron imaging facility (not to scale); the whole system is surrounded by a radiation tight shielding (mostly concrete) with permission by the IAEA [3].

a rotation stage, mainly used in tomography experiments. The precision in sample manipulation should be better than 0.1 mm depending on the detailed requirements.

The detection system (see below for more details) is the most important component in an experiment, but not necessarily the most expensive. Here, the requirement to surround the whole setup with a tight shielding comes in. Because not only the direct beam, but also scattered radiation has to be enveloped safely during beam applications, the demands for shielding are high. With the common flux level mentioned before, concrete walls of about 1-m thickness in all directions are required. In the forward beam direction, a neutron-absorbing beam catcher is additionally needed behind sample and detector.

Many other installation components are added according to the methodical requirements on demand (see Section 7).

5. Detector options

As mentioned before, camera-based detection systems are quite common in neutron imaging today. The outdated film technology with a neutron-absorbing converter was used for nearly half the twentieth century. Its advantage was high spatial resolution (about 50 μm) for relative large fields of view (FOVs) (about 30 ×30 cm). However, the acquisition time per image is on the order of 30 minutes (exposure, handling, development, fixation, and drying) at usual flux levels. This is not competitive today at all. Furthermore, film methods are not linear in their response, have low dynamic range, and cannot be used for precise quantification.

The abovementioned camera detector technology has several degrees of freedom: number of pixels, spectral range, efficiency, pixel size, and dynamic, noise behavior. In

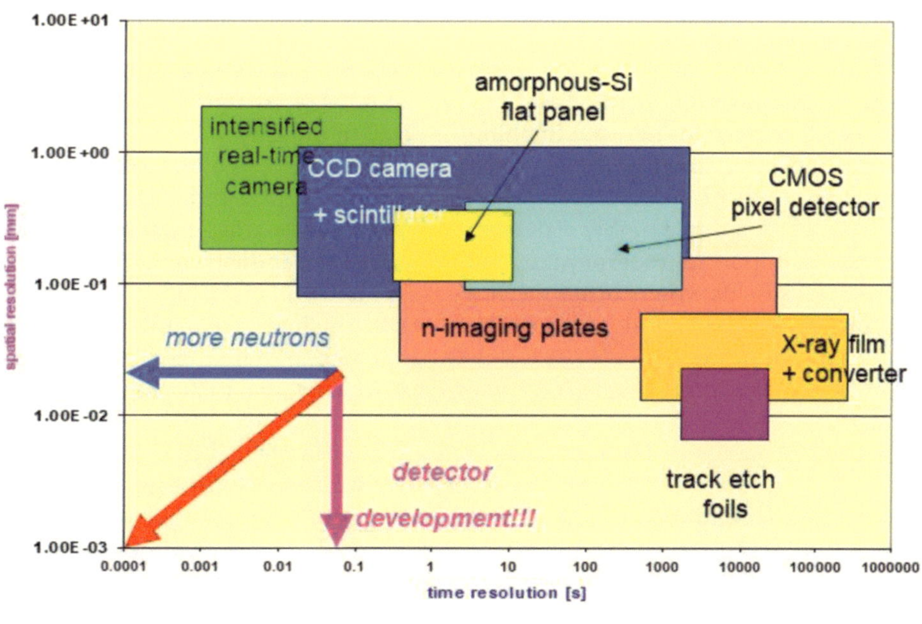

Figure 7.
Detectors for neutron imaging with their approximate working range w.r.t. spatial and time resolution, covering several orders of magnitude with permission by the IAEA [3].

addition, also the optical coupling system is relevant for the performance. Because the number of photons from the scintillator is all times limited, lenses with small f-numbers are preferred, which makes a trade-off with geometric aberrations necessary.

Some more detector options are given in **Figure 7**: flat panels based on amorphous silicon or CMOS technology, pixel sensors with absorber doped microchannels, and intensified camera systems. Recent developments with camera sensors of highest readout capability allow already counting the neutrons event by event [4].

With this high flexibility and variability, the operators of neutron imaging facilities (and their users) have to decide what are the most useful setups for their specific study. Generally, neutron imaging is limited by the number of neutrons. More neutrons would allow for higher frame rates or higher spatial resolution or better counting statistics—but never all these things at the same moment.

6. User lab function in a global context

Large-scale neutron sources (research reactors and spallation neutron sources) have been built as "user labs" with a worldwide access to beam time at high-level performing neutron research instruments, mainly for neutron scattering. Because the counting statistics at such devices is often limited in the neutron beam intensity, high source power is required.

The reactor at the Laue-Langevin Institute (ILL) in Grenoble, France, can be seen as a template for such kind of neutron source facilities. It has been functional since 1971; it is financed by 14 European countries and provides 40 neutron research instruments. The access to these instruments is organized and formalized with research proposals and supervised by scientific boards. Recently, also a neutron imaging facility (NEXT) has been installed—about 50 years after the reactor's start, indicating the increasing importance of neutron imaging techniques in "neutrons for society" (according to ILL marketing).

In the meantime, many other institutes followed the ILL scheme with their sources, often nationally operated and organized. A list of currently available user lab sources (with options for neutron imaging) is given in the Appendix.

For a while, until about 2005, neutron imaging was seen to be a simple technique without high scientific value. However, using modern imaging detection systems on the basis of digital and highly efficient methods, neutron imaging is now widely considered an important tool for many scientific, but also industrial relevant applications in various fields, where other methods are limited and often fail.

Because the experimental infrastructure for neutron imaging on high technical level can become quite expensive, it is a clever approach to share these devices also in the user lab scheme, following that for neutron scattering, and including the access rules.

High-performance neutron sources are unique with respect to their specific layout and particular performance at diverse beam lines. For neutron imaging, however, thermal and cold neutrons are the most common options; only a few facilities employing fast or epithermal neutrons are in use. No wonder, also the respective beam lines for neutron imaging have been built individually. Owing to the user lab function, a high flexibility in the infrastructure with respect to beam size, collimations, intensity, and spectrum is required. The facility ICON at the spallation neutron source SINQ, Paul Scherrer Institut (PSI), Switzerland, is shown in **Figure 8** as an example. The neutron spectrum at ICON is given in

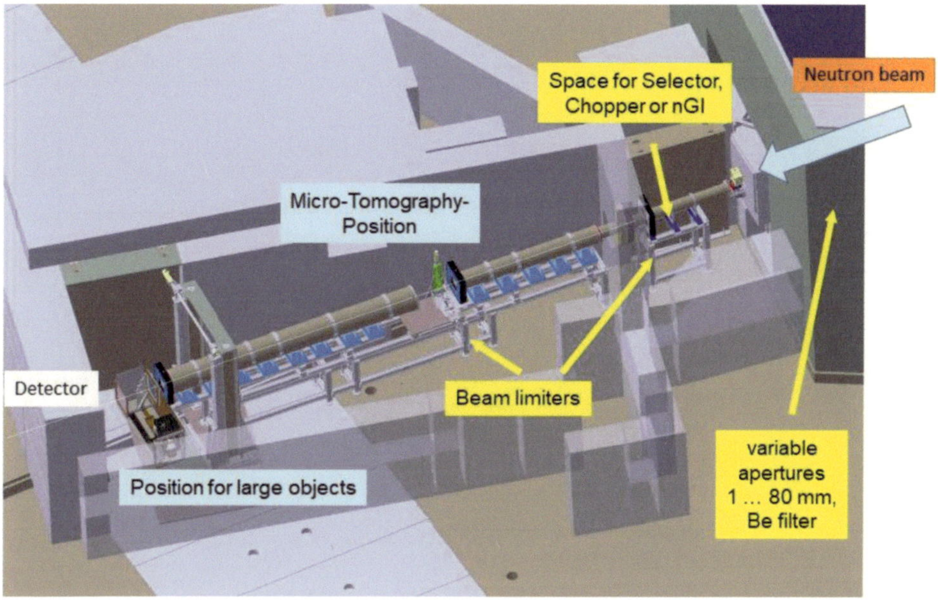

Figure 8.
Layout of the ICON neutron imaging facility for cold neutrons at PSI with the description of major installation features.

Figure 9 in comparison to the other two beams at SINQ in use for neutron imaging—demonstrating the overall high flexibility with respect to penetration power and detection sensitivity.

An overview of the current neutron imaging options and modalities was already given in **Figure 2**—accordingly, special devices and setups need to be available. However, not all sources provide the full set of options or the same conditions. Any potential applicants have to check, which facility fits best to their demands. A pretest at one or more facilities might be useful for developing a perfect setup for a specific request.

7. Methodical features in neutron imaging

The overview of "standard" and "advanced" methods has been already given by **Figure 2**. Here, we want to spend some more effort on describing particular features and their advantages for the research with neutrons.

Most of the different methods are based on the transmission image as sketched in **Figure 1**. Already this "radiography" setup has the degree of freedom for larger or smaller fields of view (FOVs), depending on the sample size, or higher or coarser spatial resolution. Because camera-based detection systems are coupled to the primary sensor (scintillator screen), the relation between FOV and pixel size can be managed by variable lens systems. The moment best layout with respect to high spatial resolution [5] enables a pixel size of 2 μm, FOV = 5×5 mm. On the other hand, the large FOV is limited by the neutron beam dimensions to about 40 ×40 cm. Even larger (but flat, e.g., paintings) objects can be studied in a scanning mode, taking images at different position by moving the sample and composing them together.

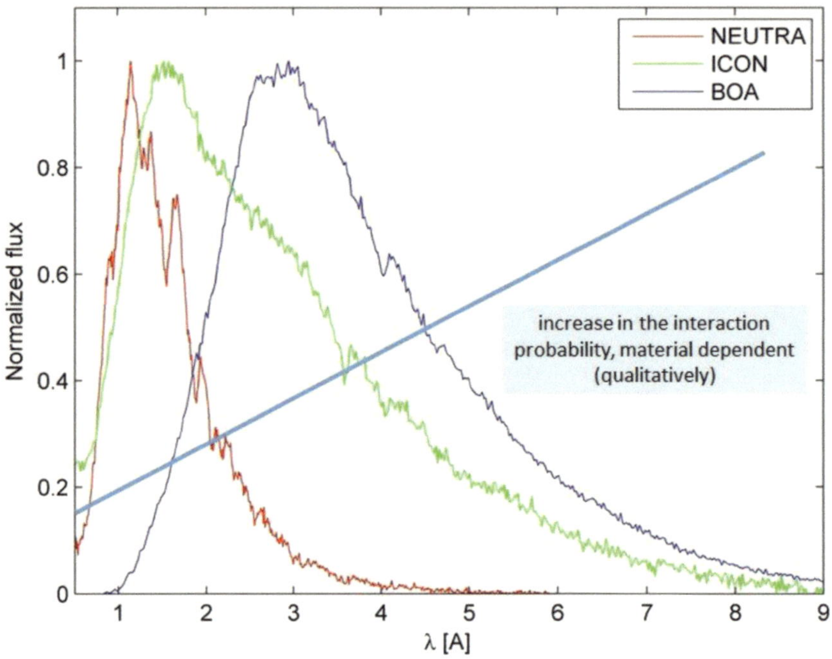

Figure 9.
Normalized neutron spectra at the PSI imaging facilities NEUTRA, ICON, and BOA and the slope of the interaction probability with Al, rising toward longer wavelengths (lower neutron energies) with permission by the IAEA [3].

Neutron tomography allows one to study samples also in their third dimension. The method works similarly to X-ray tomography in hospitals. However, neutron sources cannot be moved around the object: the scanning from different directions has to be done by sample rotation around a vertical (or horizontal) axis. A number of "projections" are taken over an angular range of at least 180° in sequential steps. They are used for the volume reconstruction harnessing mathematical procedures like "filtered back-projection" in the inverse Fourier space. The 3D neutron image of the investigated object is built from a voxel matrix of attenuation coefficients. With the help of visualization tools, slices through the object at arbitrary positions can be calculated and displayed. Regions with the same voxel value can be segmented and measured. The example in **Figure 4** compares one two-dimensional neutron radiography image of a Tibetan bronze sculpture with its 3D tomography reconstruction, for this image, cut in the middle.

Real-time image sequences can be obtained when the detector is able to provide high frame rates and corresponding read-out features. It is the advantage of modern detector options (see Section 5) to have such capabilities, at least at facilities with the required beam intensity.

Stroboscopic imaging is most useful for the observation of repetitive processes like engines. In this way, even fast motions can be investigated by a synchronization of the detection system with the process itself. Snapshots with durations of milliseconds are obtained and stacked to a valid image with good statistical accuracy. A frequency of about 800 rpm has been studied successfully [6].

Energy-selective neutron imaging has its relevance in particular for the study of crystalline solid materials like metals. Their elastic scattering behavior with neutrons is given via the Bragg condition:

$$\mathbf{n} \cdot \lambda = \mathbf{2d} \cdot \sin\Theta \tag{1}$$

with the neutron's wavelength λ, the lattice distance d, and the scattering angle θ. By choosing the right neutron wavelength λ (corresponding to the neutron energy via the de Broglie relation), crystallites with the same orientation θ can be visualized on the macroscale. In this way, textures and their modifications can be seen and analyzed even for large structures. The example in **Figure 10** shows the texture of an Al sample near a polycrystalline weld.

Because the initial beam at a neutron imaging facility is spread over a wide energy range (Maxwellian distribution around the mean thermal (25 meV) or cold (3 meV) energy), such investigations are only possible with narrowed energy bands. There are three techniques in use to select neutrons for specific energies: turbines with tilted absorber blades, double-crystal settings, and choppers for time-of-flight (TOF) options. In all the cases, the amount of the usable neutrons is strongly reduced, and the acquisition time has to be extended accordingly.

Neutron grating interferometry is based on the coherence properties in the interaction of neutrons with matter. In the understanding of neutrons as waves according to the de Broglie's relation, next to their amplitude, a phase can also be attributed [7]. If coherent neutrons with the same phase interact with samples of suitable structure, a phase shift in different regions of the sample occurs, which can be analyzed with a grating setup [8]. With this technique, magnetic domain walls can be studied online during the magnetization process [9]. An extension of this method is the determination of the "dark field image" of a sample with structures in the μm to nm range, where small-angle scattering happens [10].

Figure 10.
Structural analysis by neutron imaging of a rolled Al sample near a weld using neutrons with different wavelength; at long wavelengths (5.8 Å) and in the polychromatic beam, no texture is visible, contrary to at other wavelengths.

Diffraction imaging is based on the detection of scattered neutrons to the side of a sample by means of detection systems around it. The transmitted beam is only used as a monitor to see where the initial beam hits the sample. This method is useful mainly for samples with large grain structures and for the characterization of single crystals.

A simple setup consists of a second imaging detector perpendicularly arranged to the transmission detector at the side of the sample. During sample rotation, regions of the single crystal are illuminated with neutrons, which fulfill the condition (1) and the out-scattered signal is registered [11]. As a practical case, turbine blades of high-performance engines can be studied for their homogeneity in the structure nondestructively. For samples with smaller crystallites, multiple spots are detected by a setup covering the forward and backward regions around the sample. If the sample is rotated, many diffraction patterns can be used to derive crystalline orientation and crystallite size [12].

Imaging with polarized neutrons is based on the neutron's property to carry a magnetic moment as particle. It is oriented antiparallel to the spin vector and has a value of about $-1.9\ \mu_n$, the nuclear magneton. Because of the spin value ½, two states (up and down) are possible. In a nonpolarized beam, the two states exist equally. However, it is possible to sort out one of the states with special absorbing filters [13], and the remaining state yields a "polarized beam."

When this beam hits samples with magnetic properties, the polarization of the neutrons can be changed. An analyzer device behind the sample can observe how many polarized neutrons have changed their orientation by the sample or by the magnetic fields on their way to the detector. Therefore, imaging with polarized neutrons is very useful to investigate magnets (e.g., under superconducting conditions at low temperatures) and also magnetic fields in a certain strength range.

Data fusion with X-ray images becomes possible because common digital detection systems are able to detect both radiations in pixelwise precision. However, since the interaction with matter of neutrons and X-rays, respectively, are different in principle, two independent image data sets are produced. While X-rays only interact with the electrons in the atomic shell, neutrons "ignore" all electrons and interact with only the atomic nuclei. The higher the mass number of the atoms and the number of electrons, the higher is the X-ray contrast. This is not the case for neutrons at all—the contrast of neutron interactions depends on the peculiarities of the structure of the involved atomic nuclei. Even very light isotopes like ^{1}H, ^{10}B, or ^{6}Li have a high attenuation power, cases in which additional neutrons can readily be incorporated in the respective nuclei; in addition, in simple billiard-ball physics, energy transfer is much higher when neutrons scatter with light nuclei.

Now, it is a question how to combine the two independently obtained data sets in order to enhance features and structures. A symbolic case is the study of moisture migration in porous media (stone, concrete, wood, soil, etc.). Here, the X-rays can probe the empty structure, while the neutron image is dominated by the water contrast [13].

8. Applications in neutron imaging

Because neutrons can penetrate heavy materials, in particular metals, better than X-rays, technical applications have a high importance. This enables many industrially relevant studies like of the fuel injection in running engines, the oil distribution in combustion motors, soot accumulation in particulate filters, water distributions in electrical fuel cells, and ion migration in batteries.

The quality assurance of explosives for civil applications (initiators in rockets and detonators in mining fields) can only be done with neutron imaging.

Also, the investigation of cultural heritage objects can take advantage of neutron imaging techniques, in particular when metallic samples cover some hidden organic material—see **Figure 4** [14]. This holds also for checking the corrosion status of old samples, including the monitoring of the success of anticorrosion treatments.

Scientific usage of neutron imaging methods is widely spread. It covers different fields, listed here without going into much detail. This chapter should only give an impression how powerful and important these techniques are.

- Geoscience: structural details, porosity, organic inclusions, and water permeability [15]
- Archaeology: status of find objects, conservation, and corrosion status
- Material testing: crack analysis, connections by welding, soldering, or gluing
- Nuclear materials: fuel integrity, cladding behavior, and hydrogen uptake
- Soil and plant physics: growing of roots, water uptake, and poison analysis
- Wood research: water absorption, annual ring density, and gluing connections
- Agriculture and Food studies: fertilizer distribution, freeze drying, and moisture
- Electrochemistry: fuel cells, batteries, electrolyzes, and agent distribution
- Liquid metal behavior: bubble motion, impact of magnetic fields, and freezing
- Paleontology: bone reconstruction, specific organs, and their development
- Building material studies: moisture uptake, water isolation, and corrosion of steel

9. Conclusions: future trend and outlook

Neutron imaging depends strongly on the access to suitable neutron sources. This method alone does not justify the construction and operation of large-scale facilities as spallation sources. However, in the "concert" with other important techniques like neutron scattering and fundamental research of the neutron itself (lifetime, electrical moment), neutron imaging with all its versatile features can play an important role.

In particular, the pulsed beam option of spallation neutron sources will enable a flexible and very efficient resolution of the neutron energy in the TOF operation of beam and detection systems. The direct combination with a diffractometer device is already considered at some places [16].

However, with lower beam intensities, we can perform some neutron imaging experiments. This has been demonstrated at weak sources operating in the few watts power range [17]. New sources on the basis of proton or deuteron accelerators are under consideration and are designed in countries where a renovation of research reactors is nearly impossible, mainly for political and ideological reasons [18].

Because of their uniqueness, neutrons will be used for investigations also in near, maybe also far future, as long as the neutron sources are available. In the meantime, further methodical progress is expected, and improvements in the detection methods, like single event counting, become available.

Acknowledgements

The author thanks all involved colleagues of his team at the Paul Scherrer Institut within the Laboratory for Neutron Scattering & Imaging for their valuable contributions to the development, implementation, and usage of all described neutron imaging techniques.

A. Appendix

Neutron imaging facilities at large-scale neutron sources (Status Jan. 2023).

Institution	Country	Neutron source	Source type	Operation mode	Name of the facility	Neutron spectrum
TU Munich	D	FRM-2	reactor	stationary	ANTARES	cold/thermal
TU Munich	D	FRM-2	reactor	stationary	NECTAR	Fast/gamma
PSI	CH	SINQ	Spallation source	stationary	NEUTRA	Thermal
PSI	CH	SINQ	Spallation source	stationary	ICON	cold/thermal
PSI	CH	SINQ	Spallation source	stationary	BOA	Cold
JPARC	JP	MLF	Spallation source	Pulsed 25 Hz	RADEN	Cold
NIST	USA	NCNR	reactor	stationary	BT-2	Thermal
NIST	USA	NCNR	reactor	stationary	NG-6	Cold
ILL	EU, F	ILL reactor	reactor	stationary	NEXT	Cold
ANSTO	AUS	OPAL	reactor	stationary	DINGO	Thermal
ESS	SWE, EU	ESS	Spallation source	Pulsed 14 Hz	ODIN	Cold
ORNL	USA	HFIR	reactor	stationary	CG-9	Cold
ORNL	USA	SNS	Spallation source	Pulsed 60 Hz	VENUS	Cold
Rutherford Lab	UK	ISIS	Spallation source	Pulsed 10 Hz	IMAT	Cold
KFKI	H	WWS-M	reactor	Stationary	NORMA	Cold
KFKI	H	WWS-M	reactor	Stationary	NRAD	Thermal
JINR	RUS	IBR-2	Pulsed reactor	Pulsed 5 Hz	NR	Thermal
NECSA	ZA	SAFARI-1	reactor	stationary	SANRAD	Thermal

Author details

Eberhard H. Lehmann
Paul Scherrer Institut, Switzerland

*Address all correspondence to: eberhard.lehmann@psi.ch

References

[1] Kalmann H. Research. 1947;**1**:254-260

[2] Peter OZ. Naturforschung A. 1946;**1**:557-559

[3] Lehmann E. Status and Progress in Neutron Imaging Detection Systems. IAEA-TECDOC. Vienna: IAEA; 2020. pp. 261-271

[4] Losko A. New perspectives for neutron imaging through advanced event-mode data acquisition detector. Scientific Reports. 2021;**11**(1):21360. DOI: 10.1038/s41598-021-00822-5

[5] Trtik P, Lehmann E. Progress in high-resolution neutron imaging at the Paul Scherrer Institut - the neutron microscope project. Journal of Physics: Conference Series. 2016;**746**:012004

[6] Grünzweig C. Progress in industrial applications using modern neutron imaging techniques. Physics Procedia. 2013;**43**:231-242

[7] Pfeiffer F. Milestones and basic principles of grating-based x-ray and neutron phase-contrast imaging. AIP Conference Proceedings. 2012;**1466**:2. DOI: 10.1063/1.4742261

[8] Grünzweig C et al. Design, fabrication, and characterization of diffraction gratings for neutron phase contrast imaging. Review of Scientific Instruments. 2008;**79**:053703. DOI: 10.1063/1.2930866

[9] Betz B, Grünzweig C, Lehmann E. Advances in neutron imaging with grating interferometry. Materials Evaluation. 2014;**72**:491-496

[10] Strobl M et al. Physical Review Letters. 2008;**101**:123902

[11] Petermans S, Lehmann E. Simultaneous neutron transmission and diffraction contrast tomography as a non-destructive 3D-method for bulk single crystal quality investigations. Journal of Applied Physics. 2013;**114**:124905

[12] Samothrakitis S et al. Laue 3-dimensional neutron diffraction tomography and the FALCON instrument. Swiss Neutron News No. 2022;**60**:6-18

[13] Treimer W et al. Polarized neutron imaging and three-dimensional calculation of magnetic flux trapping in bulk of superconductors. Physical Review B. 2012;**85**:184522

[14] Mannes D, Lehmann E. Neutron imaging of cultural heritage objects. In: D'Amico S, Venuti V, editors. Handbook of Cultural Heritage Analysis. Switzerland AG: Springer Nature; 2022. DOI: 10.1007/978-3-030-60016-7_9

[15] Tengatini A et al. Neutron imaging for geomechanics: A review. Geomechanics for Energy and the Environment. 2021;**27**:100206

[16] Kockelmann W et al. IMAT–A new imaging and diffraction instrument at ISIS. Physics Procedia. 2013;**43**:100-110

[17] Lange C, Berndt N. Neutron imaging at the low flux training and research reactor AKR-2. NIM A. 2019;**941**:162292

[18] Brückel T, Gutberlet T, editors. Conceptual Design Report. Jülich High Brilliance Neutron Source (HBS), Forschungszentrum Jülich: Germany; 2020